Transactions on Computational Science and Computational Intelligence

Series Editor
Hamid Arabnia

Computational Science (CS) and Computational Intelligence (CI) both share the same objective: finding solutions to difficult problems. However, the methods to the solutions are different. The main objective of this book series, "Transactions on Computational Science and Computational Intelligence", is to facilitate increased opportunities for cross-fertilization across CS and CI. This book series will publish monographs, professional books, contributed volumes, and textbooks in Computational Science and Computational Intelligence. Book proposals are solicited for consideration in all topics in CS and CI including, but not limited to, Pattern recognition applications; Machine vision; Brain-machine interface; Embodied robotics; Biometrics; Computational biology; Bioinformatics; Image and signal processing; Information mining and forecasting; Sensor networks; Information processing; Internet and multimedia; DNA computing; Machine learning applications; Multi-agent systems applications; Telecommunications; Transportation systems; Intrusion detection and fault diagnosis; Game technologies; Material sciences; Space, weather, climate systems, and global changes; Computational ocean and earth sciences; Combustion system simulation; Computational chemistry and biochemistry; Computational physics; Medical applications; Transportation systems and simulations; Structural engineering; Computational electro-magnetic; Computer graphics and multimedia; Face recognition; Semiconductor technology, electronic circuits, and system design; Dynamic systems; Computational finance; Information mining and applications; Astrophysics; Biometric modeling; Geology and geophysics; Nuclear physics; Computational journalism; Geographical Information Systems (GIS) and remote sensing; Military and defense related applications; Ubiquitous computing; Virtual reality; Agent-based modeling; Computational psychometrics; Affective computing; Computational economics; Computational statistics; and Emerging applications. For further information, please contact Mary James, Senior Editor, Springer, mary.james@springer.com.

More information about this series at http://www.springer.com/series/11769

Fadi Al-Turjman
Editor

Artificial Intelligence in IoT

Springer

Editor
Fadi Al-Turjman
Computer Engineering Department
Antalya Bilim University
Antalya, Turkey

ISSN 2569-7072 ISSN 2569-7080 (electronic)
Transactions on Computational Science and Computational Intelligence
ISBN 978-3-030-04109-0 ISBN 978-3-030-04110-6 (eBook)
https://doi.org/10.1007/978-3-030-04110-6

Library of Congress Control Number: 2019932816

This Springer imprint is published by the registered company Springer Nature Switzerland AG.
The registered company address is: Gewerbestrasse 11, 6330 Cham, Switzerland

To my wonderful family.

Fadi Al-Turjman

Preface

We are living in an era where the Artificial Intelligence (AI) is becoming a global platform for the computation and interaction between machines and smart objects in real-time applications and many critical aspects in our daily life.

With the application areas such as smart cities, smart grids, smart water managements, smart health, smart supply chain, and smart homes in the Internet of Things (IoT), we can consider the AI as a complementary package of the smart networked objects. From this perspective, it is essential to understand the role of AI which will provide a global backbone for the worldwide information sharing/processing in the near future.

No doubt that introducing such a new phenomenon can come with potential challenges in significant levels, especially in terms of the overall system complexity and ability in solving critical daily life issues. Therefore, it is also essential to consider new enabling technologies such as wireless sensor networks (WSNs), various radio technologies, and cellular infrastructures for the performance optimization.

The objective of this book is to present a survey of existing AI techniques and other emerging intelligent approaches for the IoT paradigm optimization and improvements. The main focus is on the smart design aspects that can help in realizing such a paradigm in an efficient and secured way. The applications of AI in IoT, evaluation metrics, constraints, and open issues about the addressed topics are included for discussion as well. This conceptual book, which is unique in the field, will assist researchers and professionals working in the area of AI to better assess the proposed IoT paradigms which are already beginning to be a significant part of the global infrastructure on the planet.

Hope you enjoy it.

Antalya, Turkey
Fadi Al-Turjman

Contents

About the Editor

Fadi Al-Turjman is a Professor at Antalya Bilim University, Turkey. He received his Ph.D. degree in computing science from Queen's University, Canada, in 2011. He is a leading authority in the areas of smart/cognitive, wireless, and mobile networks' architectures, protocols, deployments, and performance evaluation. His record spans over 180 publications in journals, conferences, patents, books, and book chapters, in addition to numerous keynotes and plenary talks at flagship venues. He has authored/edited more than 12 published books about cognition, security, and wireless sensor networks' deployments in smart environments with Taylor & Francis and Springer (top-tier publishers in the area). He was a recipient of several recognitions and best papers awards at top international conferences. He led a number of international symposia and workshops in flagship ComSoc conferences. He is serving as the Lead Guest Editor in several journals, including the IET Wireless Sensor Systems and Sensors (MDPI and Wiley). He is also the Publication Chair for the IEEE International Conference on Local Computer Networks.

A Systematic Review of the Convergence of Augmented Reality, Intelligent Virtual Agents, and the Internet of Things

Nahal Norouzi, Gerd Bruder, Brandon Belna, Stefanie Mutter, Damla Turgut, and Greg Welch

1 Introduction

In a seminal article on *augmented reality* (AR) [7], Ron Azuma defines AR as a variation of virtual reality (VR), which completely immerses a user inside a synthetic environment. Azuma says "In contrast, AR allows the user to *see the real world*, with virtual objects *superimposed upon or composited with the real world*" [7] (emphasis added). Typically, a user wears a tracked stereoscopic head-mounted display (HMD) or holds a smartphone, showing the real world through optical or video means, with superimposed graphics that provide the appearance of virtual content that is related to and registered with the real world. While AR has been around since the 1960s [72], it is experiencing a renaissance of development and consumer interest. With exciting products from Microsoft (HoloLens), Metavision (Meta 2), and others; Apple's AR Developer's Kit (ARKit); and well-funded startups like Magic Leap [54], the future is looking even brighter, expecting that AR technologies will be absorbed into our daily lives and have a strong influence on our society in the foreseeable future.

At the same time, we are seeing the continued evolution of *intelligent virtual agents* (IVAs) in the home through products such as Apple's Home Pod, Amazon's Echo, and Google Home. Gartner predicted that the IVA market will reach $2.1 billion by 2020 [58]. The products use sophisticated microphones and signal

N. Norouzi (✉) · G. Bruder · B. Belna · D. Turgut · G. Welch
University of Central Florida, Orlando, FL, USA
e-mail: nahal.norouzi@knights.ucf.edu; bruder@ucf.edu; bbelna@knights.ucf.edu; turgut@cs.ucf.edu; welch@ucf.edu

S. Mutter
Muhlenberg College, Allentown, PA, USA
e-mail: stefaniemutter@muhlenberg.edu

F. Al-Turjman (ed.), *Artificial Intelligence in IoT*,
Transactions on Computational Science and Computational Intelligence,
https://doi.org/10.1007/978-3-030-04110-6_1

processing to capture human speech in one or more rooms of one's house and via artificial intelligence algorithms and Internet-based services play music, answer basic questions, and check on sports scores, weather, and more. Products like the Gatebox [27], a Japanese take on Amazon's Echo with a holographic display, even provide an embodied representation of the IVA for users to interact with. Various research prototypes in the field of IVA used projectors and TV screens to study natural interaction with embodied virtual agents in the past. While research on IVAs started as an independent trend from AR, we are now seeing that over the last years, more and more technologies and techniques from AR are used for IVA research and commercial developments and vice versa.

The abilities of these products further extend to home automation and more general interactions with the increasingly present *Internet of things* (IoT) [1], i.e., a network of sensors and actuators within or attached to real-world objects. The term was coined by Kevin Ashton [5], who at MIT's AutoID lab in the early 2000s was laying the groundwork for what would become IoT. Cisco's IoT Group predicts there will be over 500 billion connected devices by 2030 [79]. Amazon and other companies networked their IVA devices to IoT and related smart home appliances and found an important application field for IVAs, resulting in a novel research thrust and mutually beneficial overlap between these fields. Many of the research topics pursued in the IoT field, such as privacy [56, 73], the relationship between edge and cloud processing [15], and network traffic optimization [2], will need to be re-evaluated when IoT is deployed in the context of AR and IVA. Furthermore, some IoT applications, such as smart healthcare [3, 77], can benefit from the addition of AR and IVA techniques.

Research in the three fields AR, IVA, and IoT has led to a large body of literature, and active research communities that traditionally received limited input from each other, advancing knowledge and developing novel technologies and systems within each field individually (see Fig. 1). However, over the last decades, we have seen an increasing integration of knowledge, technologies, and methods from

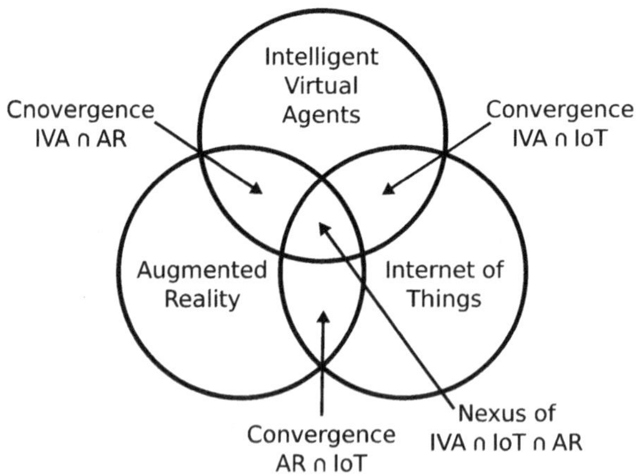

Fig. 1 Venn diagram illustrating the convergence of augmented reality, intelligent virtual agents, and the Internet of things

the three fields that act as frameworks to catalyze new research and development. The National Science Foundation (NSF) acknowledged the importance of such a convergence of research fields as one of the *10 Big Ideas for Future NSF Investments* [30]. According to NSF, "convergence research" is closely related to *transdisciplinary* research, which is generally viewed as the pinnacle of evolutionary integration across disciplines. However, convergence research represents more than transdisciplinary, interdisciplinary, and multidisciplinary research in that the fields not only overlap and integrate but come together to establish a new field with new directions for research that can attract and draw from deep integration of researchers from different disciplines, leveraging their collective strengths, with the hope of solving new or vexing research challenges and opportunities.

In this chapter, we present a review of 187 publications scattered throughout scientific conferences and workshops in diverse research communities over the last decades at the intersections of each two of the three fields AR, IVA, and IoT. We identified impactful research papers and directions for research in these fields, and we discuss a vision for the nexus of all three technologies. We highlight key themes and identify possible future research topics and trends. Overall, providing a review that introduces a substantial and useful perspective, focusing on the era of convergence research in these areas is the main goal. We hope that this paper can benefit new researchers and students involved in academia by providing a summary of the current advances and trends in these research areas and help them identify their research interests. We also hope that it will be helpful for senior researchers to see a big picture of AR, IVA, and IoT research trends, particularly with respect to a future vision of AR that may have a positive influence on humans and our society.

The remainder of this paper is structured as follows. We first describe our review methodology in Sect. 2. Then, we present a high-level description of our considered review topics in Sect. 3, which is followed by a meta-review of publications on AR, IVA, and IoT in Sect. 4. We then present a review of existing convergent research on AR and IVA, IVA and IoT, and AR and IoT in Sects. 5, 6, and 7, respectively, and we discuss trends that were observed from our reviews. In Sect. 8, we discuss a novel vision for future research that we see at the nexus of IVA, IoT, and AR. We conclude the paper in Sect. 9.

2 Methodology

For our literature review, we were faced with the challenge that papers published at the intersections of AR, IVA, and IoT could be published in a wide range of research communities with their own journals, conferences, and workshops. We could further not rely on an established terminology that would describe the convergence research and could guide our literature search. We decided on the following two-tailed literature search strategy to locate relevant publications:

1. We conducted a computerized search for publications in the online digital libraries of the *Association for Computing Machinery (ACM)*, *Institute of Electrical and Electronics Engineers (IEEE)*, *Eurographics*, *Elsevier*, and *Google Scholar* databases. Searches were conducted using a 132-item list of relevant terms, such as "augmented reality," "mixed reality," the "Internet of things," "smart home," etc. with each requiring combinations of two of these terms to identify publications in the intersections of the respective fields. The terms in each area were searched in the title, abstract, and keyword fields of the above libraries if available.
2. We searched the reference lists of located relevant publications for further relevant literature.

The abstract and body of each located publication was examined, and each was selected for further analysis if and only if it matched all of the following criteria:

1. The publication was peer reviewed and published at a scientific journal, conference, and workshop or as a book chapter. Technical reports, posters, and demos were not considered since they are usually shorter and/or not normally reviewed as rigorously.
2. The publication was at least four pages long. Shorter publications were excluded to limit the search to mature research and avoid position papers and work in progress.
3. The publication was released in the new millennium, i.e., in the year 2000 or later, to limit the scope of the literature review to a tractable period.
4. We consider certain related topics as outside of the scope of this literature review:

   (a) We excluded publications on intelligent agent software algorithms that had neither a 2D/3D virtual representation nor a voice-based natural user interface.
   (b) We excluded agents with physical manifestations as robotic humans.
   (c) We excluded VR and "augmented virtuality" display environments [57].
   (d) We excluded wearable devices like smart fabrics, wrist, belt, or foot-worn devices, if they were not connected to the Internet.

From now on we use the mathematical intersection operator, e.g., AR $\cap$ IVA, to indicate the set of publications that include both AR and IVA concepts and satisfied the above criteria, i.e., publications at the intersection of the respective fields.

The above procedures resulted in a sample of 187 publications in total between the years 2000 and 2018 for the fields AR $\cap$ IVA (65 publications), IVA $\cap$ IoT (43 publications), and AR $\cap$ IoT (79 publications). Of course, we do not make any claims that this list of publications covers the entirety of research in the identified converging fields, but we hope that the analysis and review of the located publications can provide an excellent snapshot of the work listed at these premier tools for disseminating academic research.

The second part of the review process focused on the following points:

1. We divided the total number of publications among the first four authors and classified all publications based on their research contributions using the research topics described below.
2. We collected the citation counts for all publications on August 19, 2018. Due to the different sources of the publications, we decided to collect citation counts from *Google Citation Index*, which covered the largest portion of the publications compared to *Scopus* and other databases. If we could not find a citation count for a specific publication on any of the databases, we set it to zero for the analysis.
3. We divided the publications among the first four authors based on their expertise and performed in-depth reviews of the most highly cited publications in the converging fields. If publications were released recently (in 2017 or later), we did not entirely rely on the citation counts as a measure of impact but included a quick first review of the publications before performing an in-depth review. We considered analyzing further metrics of the impact of these publications, but honors and awards such as *Best Paper* or *Honorable Mention Awards* proved intractable for our diverse set of publications.

During the review process, we focused on identifying and reviewing those publications that had a high impact on the convergence of the fields, novel trends, and directions that stemmed from those publications, as well as challenges that were encountered.

3 Review Topics

During the literature search process, we identified groups of publications related to different research topics that guided our analysis process. We refined our list of topics based on related literature surveys, including [22, 41, 59]. We refined or removed topics from that list based on the number of related publications in our literature search. In the end, we decided to group all publications into seven research topics, ranging from core technology areas needed to deliver an AR, IVA, or IoT application to emerging research fields. Publications may belong to one or more of these topic categories.

The research topic categories are:

1. *System*: research on systems covering at least in part two of the three areas AR, IVA, and IoT
2. *Application*: research on systems in application domains such as manufacturing, healthcare, and defense, among others
3. *Evaluation*: research focusing on human-subject studies evaluating systems or techniques
4. *Review/Survey*: literature reviews including at least in part two of the considered convergence areas
5. *Multimodal*: research into combined modalities such as speech and gesture interfaces

6. *Collaboration*: research on interactive collaborative systems for multiple users
7. *Interaction*: research on user interfaces or interaction techniques

We further aimed to understand the input and output modalities of these systems at the intersections of AR, IVA, and IoT in more detail. We hence decided to classify all publications based on two general concepts related to the information transfer between the real/physical world and the virtual/computer-controlled content during interactions:

- *Input*: The "virtual" is aware of something "real."
- *Output*: The "virtual" is influencing something "real."

We considered the input and output dimensions for different human sensory and related modalities: light, sound, touch, motion, temperature, fluid, smell, and airflow. Most of the publications addressed light or sound output (e.g., visual or auditory displays in AR), some considered light or motion input (e.g., camera-based user tracking in AR) and sound input (e.g., voice-controlled agents), and some considered physical output (e.g., exerting influence over smart homes with IoT). For each paper we also looked at the types of displays used for the AR content and the IVAs. The displays considered are *HMDs*, *mobile* devices (e.g., cell phones, tablets), *screens* (flat screens, TVs), and *other* less often occurring types of displays that we grouped together such as projection-based AR and CAVE environments. We used the tag *N/A* for publications where a display type was not indicated as part of their work, and we also categorized publications that developed setups that are migratable to multiple different displays as *cross-display*.

4 Meta-Review of Publications

In this section, we describe a high-level meta-review analysis of the number of publications and citations for each convergence field and research category. As stated above, we considered a total of seven research topics that we want to discuss in this chapter. First, we evaluated the number of publications for each convergence field over time (see Fig. 2). Then, we evaluated them for each research category and their percentage over the total number of classifications (see Fig. 3). Publications could cover multiple topics; thus, this classification count is larger than the number of publications—338 classifications among 187 publications.

As shown in Fig. 2, research in the field of AR ∩ IVA goes further back compared to the other two fields, which is understandable considering that IoT is a comparatively more recent area of research. Also, in line with the technological advances in the field of IoT in the past few years (e.g., Amazon's Alexa, Microsoft's Cortana, and the Google Assistant), we observe a significant increase in the number of publications in both of the fields converging with IoT. Interestingly, the drop in the number of publications in 2009 and 2010 coincides with a rise in publications in the fields of AR ∩ IoT and IVA ∩ IoT.

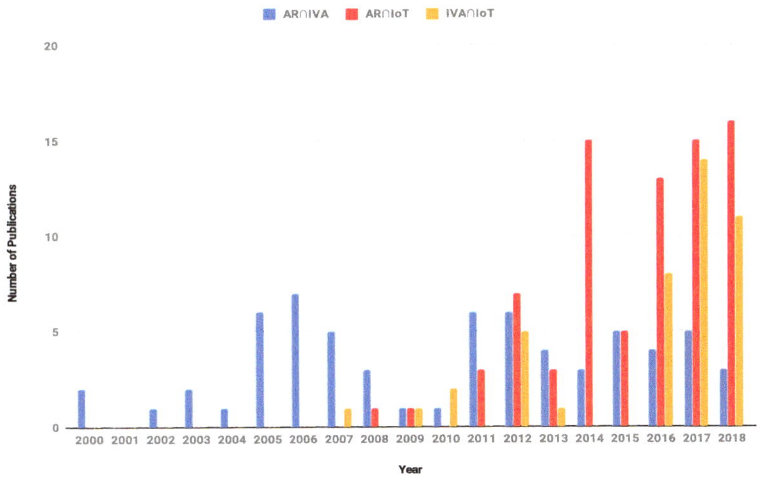

Fig. 2 Number of publications in the convergence fields from the year 2000

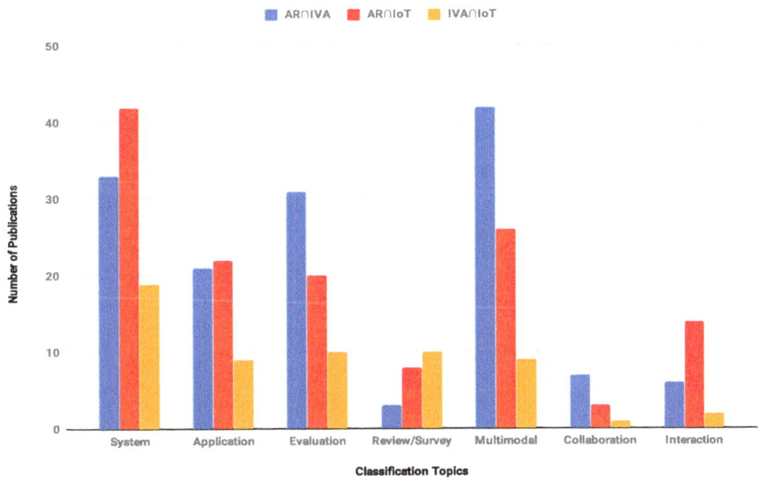

Fig. 3 Number of publications in each convergence field grouped according to classification topics

Looking at the distribution of classification topics for each field in Fig. 3, excluding review and survey papers, we observe a lower number of publications in the remaining topics for the field of IVA ∩ IoT, which is partly due to our exclusion criteria #4 in Sect. 2 and the novelty of this topic. Also, with research in the field of AR and IVA being more developed than IoT, we observe fewer system papers in AR ∩ IVA compared to AR ∩ IoT and more research involving user studies.

Interestingly, even though the research in the fields of AR and IVA goes back further than IoT, we see a similar number of application papers in AR ∩ IVA compared to AR ∩ IoT, which in part may be due to the sudden growth in using

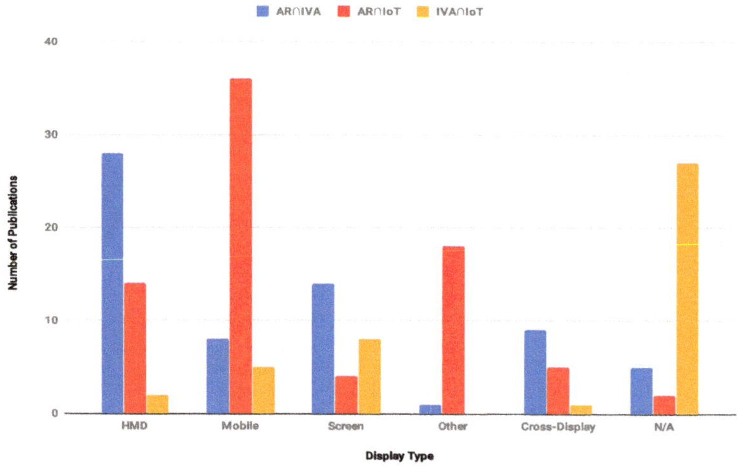

Fig. 4 Number of publications in each convergence field grouped by display type

IoT devices in one form or another by the general public unlike AR ∩ IVA, due to high price or technical difficulties of developing and interacting with holograms.

Looking at the type of displays used in each field in Fig. 4, we observe that the majority of publications in the field AR ∩ IVA used some type of HMD, either video see-through or optical see-through, compared to using mobile devices such as cell phones and tablets. The opposite pattern is observed with AR ∩ IoT. We see more publications in AR ∩ IVA that built frameworks and/or applications that were migratable to different display hardware, which can be explained by the age of this research field. Many publications especially the ones involving IoT did not necessarily mention or use a display device mostly because their research was focused on techniques, algorithm, and theoretical frameworks. We also observe more publications focusing on new interfaces and interaction techniques in AR ∩ IoT which is understandable as virtual assistants and information overlay through mobile devices has shown to be a popular topic in this field.

We also computed the average yearly citation counts (ACC) for each field of research which is shown in Fig. 5. The sudden increase in ACC for both AR ∩ IoT and IVA ∩ IoT is indicative of the fast-paced improvements and high impacts in both fields considering the high ACCs in 2016 and 2017.

5 The Convergence of AR and IVA

Although extensively researched in VR, intelligent agents have been introduced to AR for the first time at the beginning of the new millennium. This convergence field brings together the field of AR, in which *virtual* content is overlaid or mixed with

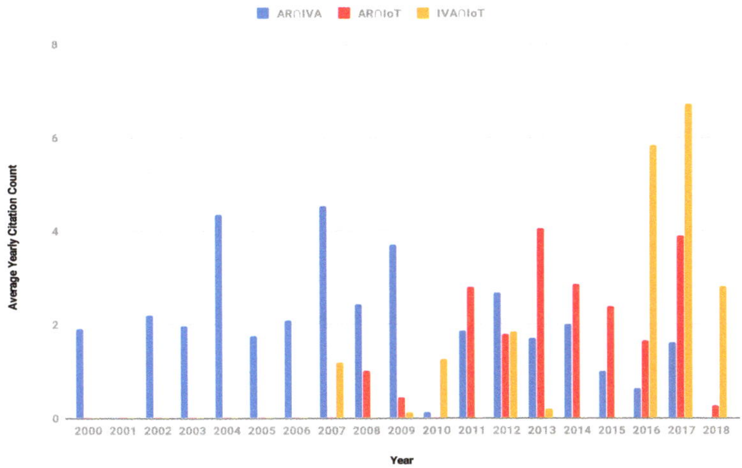

Fig. 5 Average yearly citation count per convergence field

the *physical* world, and the field of IVA, which is concerned with *virtual* agents and how these can interact with *physical* users and objects.

5.1 Meta-Review

At the intersection of AR and IVA, we found 65 publications with 143 classifications in total. Figure 2 shows the number of publications by year, and Fig. 3 shows them by classification. As shown in Fig. 3, the majority of the publications in this field included aspects of system development (33), evaluation (31), and employed agents and/or setups that supported several input/output modalities (42) with fewer papers focusing on applications (21), collaboration (7), interaction (6), and survey papers (3).

Looking at the types of displays used per year in Fig. 6, we noticed an inclination towards HMDs (28). The second most used display type were flat screens (14), e.g., computer display, TV, etc., followed by publications supporting several hardware platforms (9); mobile devices (8), e.g., cell phones and tablets; and a CAVE environment (1). Five papers did not focus on aspects that would require specific displays.

We also looked at the input/output modalities used in the publications in this field. As shown in Fig. 7, light followed by sound were employed as the main modalities of the works published. We observe a similar number of input and output publications for light since many of the devices used in this field included depth sensors and cameras as input and overlaid virtual content as output. We also see a higher number of publications using the sound modality as output due to many virtual characters being able to speak to users but not having speech recognition

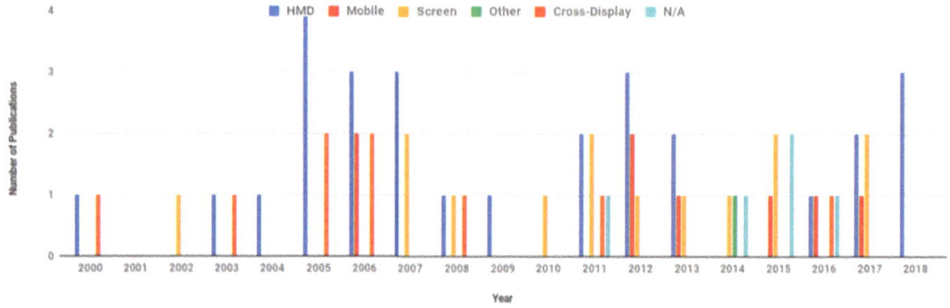

Fig. 6 Variety of displays used over time in the field of AR and IVA

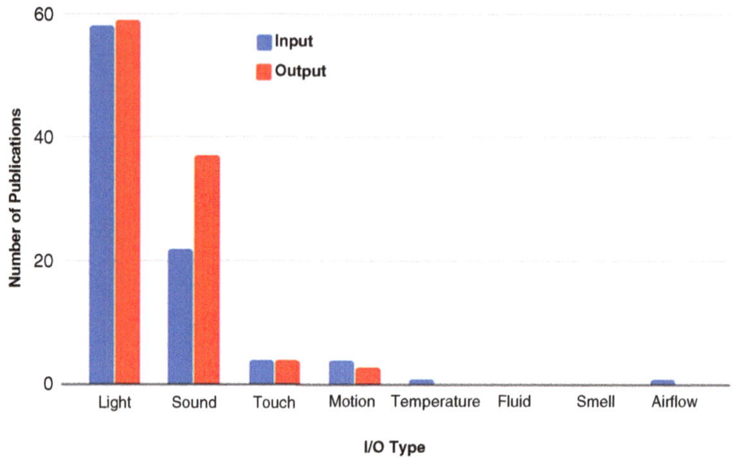

Fig. 7 Publications in the field of AR and IVA grouped based on the modalities used

capabilities. A few publications adopted sensors to bring about touch and motion capabilities and only one paper experimented with airflow.

Ten publications in this field had a citation count of four citations per year or higher. The average citation count in this field was 1.9 citations per year ($SD = 1.8$). The papers below were selected based on the citation count and qualitative criteria for years 2017 and 2018, for which the citation counts are not yet well established.

5.2 *Impactful Papers and Trends*

Barakonyi et al. [10] presented the first general framework for IVAs called *AR Puppet,* which has been developed specifically for AR applications. They discussed agent-specific aspects of AR such as that IVAs should react to the physical environment, avoid collisions with physical obstacles, and should be able to influence objects in the real world. In later work, the authors introduced numerous AR applications in this field [8, 9, 11]. For instance, they presented an AR tabletop game called MonkeyBridge [11] in which small embodied IVAs dynamically recognized changes and events in both the virtual and physical world and adapted their movement behavior to these changes. Similar responses to dynamically changing physical environments in AR for IVAs were realized by Chekhlov et al. [18] in their high-impact *Ninja on a Plane* game, with SLAM-based sensing of the physical world. Blum et al. [14] presented an interactive location-aware AR game based on mobile AR devices and IVAs, which uses real locations and includes virtual characters with basic physical behaviors. They showed in a user study that the sense of embodiment has a positive impact on player experience.

Multiple publications were focused on IVAs in AR that exhibit more sophisticated reasoning, proactive behavior, and some awareness of the non-verbal behaviors of a real human [44, 45, 52, 67]. Others examined the effects of awareness and influence of IVAs in AR settings using cameras, microphones, and custom devices to develop automated awareness of environmental effects such as sounds, movements (felt and observed), light, air flow, and IVA interactions with other humans [21, 39, 40, 48, 50, 70]. The results indicate that, like VR, congruent and plausible behaviors matter—virtual humans and objects should be responsive to real-world events. However, in practice, it is much more difficult in AR, because unlike the case with VR, where the system controls everything, AR systems are typically unaware of dynamic real people and events. As recognized by Helen Papagiannis, "the second wave of Augmented Reality (AR) is driven by a contextual understanding and interaction with your surroundings" [60].

Much high-impact research on IVAs in AR was driven by application fields demanding more useful and engaging IVA technology. For instance, Hantono et al. [31] reviewed the literature between 2005 and 2015 concerning IVAs in AR with respect to possible uses in *education*. They observed that most AR realizations at that time were not personalized to the user or the environment, limiting the learning experience. Wagner et al. [75] presented one of the first IVAs in (hand-held) AR for educational purposes. A wide range of prototypes of IVAs in AR were developed for application contexts [4, 17, 30, 42, 61].

In a highly cited paper, Dow et al. [23] described an interactive drama using autonomous characters presented on an AR HMD, artificial intelligence-based story management, and natural language processing to immerse the player in a dramatic story in AR. They showed with a qualitative study that AR interfaces make the experience with IVAs more immediate and can lead to a higher level of perceived presence in a dramatic story than desktop interfaces. Interestingly, their results also showed that this AR-enhanced immediacy of the experience does not necessarily lead to higher engagement with IVAs since players do not perceive there to be a safe zone for them to experiment and make mistakes.

Holz et al. [36] presented a multidimensional classification method to understand differences in the realization of IVA prototypes in mixed reality called the *MiRA (Mixed Reality Agents) Cube*, which divides agents based on their corporeal presence (virtual to physical), their interactive capacity (virtual to physical), and their agency (weak to strong).

While most embodied IVAs in AR are presented on projection screens or HMDs, Krum et al. [46] presented a mixed reality projection framework called *REFLCT* for IVAs and virtual content in general that couples a near-axis head-mounted projector with retroreflective props and surfaces to provide personalized, perspective-correct imagery that is uniquely composited for each user directly into and onto a surrounding environment. They demonstrated it with a virtual character that made eye contact with each person in a group of visitors.

Lee et al. [51] evaluated an optical see-through HMD as a means to present IVAs in AR. They performed a user study and showed that the limited field of view of current-state AR HMDs changes users' locomotion behavior and proxemics in the presence of an embodied IVA. They further found that subtle vibrotactile feedback of the IVA's footsteps transmitted through the floor while walking can significantly benefit a user's sense of copresence.

Overall, with a few exceptions, the highest impact in the convergence field of IVA and AR so far was made by publications focusing on system prototypes driven by application needs. Researchers have made continuous low-key progress on IVAs in AR since the early 2000s, but research on the nature and underlying peculiarities of IVAs in AR is still in its infancy compared to related efforts for IVAs in VR.

6 The Convergence of IVA and IoT

While the concept of IoT goes back to the early 2000s, it took until around 2014 with the release of Amazon Echo, and the integration of IoT devices into the agent's affordances, for research in this convergence field to take off. This convergence field brings together the field of IVA, which is concerned with *virtual* agents and how these can interact with *physical* objects and users, and the field of IoT, which strives to network *physical* devices, including sensors, appliances, and displays, to improve interaction and control by *virtualizing* the components.

6.1 Meta-Review

At the intersection of IVA and IoT, we found 43 publications with 60 classifications in total. Figure 2 shows the publications by year, and Fig. 3 shows them by classification. Fifteen publications in this field had a citation count of four citations per year or higher. The average citation count in this field was elevated to 4.29 citations per year with a large variance ($SD = 6.32$) due to a few highly cited papers in this field.

As shown in Fig. 3, the majority of the publications in this field focused on system development (19), followed by evaluation (10), survey (10), application (9),

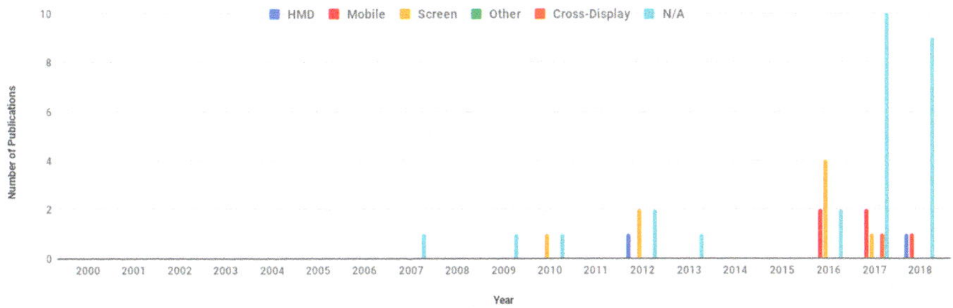

Fig. 8 Variety of displays used over time in the field of IVA and IoT

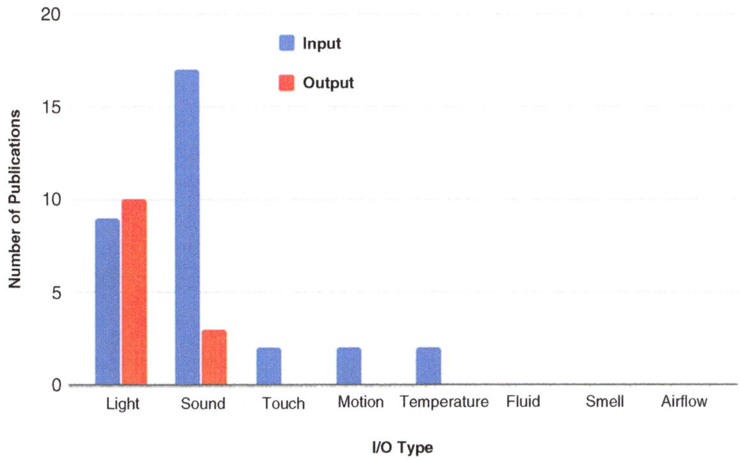

Fig. 9 Publications in the field of IVA and IoT grouped based on the modalities used

and multimodal aspects (9). Considering the novelty of the field, only one publication investigated collaborative opportunities, and two publications introduced new interaction techniques.

As shown in Fig. 8, with publications focusing on voice-based assistants such as Amazon Alexa, algorithms and theoretical frameworks, and survey papers, the majority of them did not include a specific type of display (27), followed by flat screen displays (8), handheld devices (5), HMDs (2), and one paper that focused on several hardware devices.

Looking at the type of input/output modalities employed in the field of IVA ∩ IoT, in Fig. 9 we see a similar pattern as in the field of AR ∩ IVA with regard to the emphasis on using the light and sound modalities, but the difference here is the increase in the number of publications that utilized sound as input through speech recognition which is in line with many smart home virtual assistant systems.

6.2 Impactful Papers and Trends

Soda et al. [71] developed one of the first prototypes of a voice-based IVA system in the scope of a research project, using a device with a microphone array as an interface for smart home devices, similar to the later consumer devices popularized by Amazon, Apple, and Google, among others, for natural interaction with IoT devices.

Multiple highly cited papers focused on comparative evaluations of the different consumer products being released over the last years in the field of IVA and IoT. For instance, López et al. [53] presented a comparison of existing speech-based IVAs like Amazon Alexa and Google Assistant. The authors describe a user study in which the effectiveness of the speech-based user interfaces was evaluated. Reis et al. [66] proposed a model that allows elderly people to more easily interact with IVAs, and they compared Google Assistant, Amazon Alexa, and others to evaluate which of the devices/assistants would incorporate the proposed model most efficiently. Druga et al. [24] conducted a user study that involved children interacting with IVAs including Amazon Alexa among others. The children were surveyed with regard to the agents' trust, intelligence, personality, and engagement. The authors proposed design modifications for IVAs that are capable of interacting with children in the future. Hoffman and Novak [35] provided a general discussion of the experience of consumers when interacting with IVAs and IoT networks.

Knote et al. [42] presented a recent literature review of applications for IVA assistants. The authors identified three application domains (daily life and family, work support, and e-learning) in which future research is particularly promising. Austerjost et al. [6] developed voice-based agent user interfaces for a virtual "lab assistant." The virtual assistant, through IoT devices, could control and modify physical devices and data and react to speech command inquiries.

With the aim to better understand IoT networks in smart home environments, Helal et al. [33] and Lee and Cho et al. [49] presented simulators with 3D autonomous agents that populate a smart home environment, trigger motion detectors, etc. They modeled a typical human behavior and provide a user interface to keep track of the state of the IoT network and smart environment.

Since all new technology, including IVAs and IoT in digital ecosystems like Amazon Alexa, has security concerns and could be used to commit crimes due to natural voice-based interfaces with the IVA and related vulnerabilities or could in general be important as potential sources of digital evidence for crimes committed in the real world [19], Chung and Park et al. [20] developed a proof-of-concept digital forensics tool, called *CIFT*, that supports identification, acquisition, and analysis of the state and actions within such an IVA-IoT ecosystem. To prevent the illegitimate use of one's private data in an IVA-IoT ecosystem, e.g., in commercial systems, Campagna et al. [16] developed open, crowdsourced, privacy-preserving, and programmable virtual assistant architecture, called *Almond*.

Overall, early research on the convergence of IVA and IoT had not received much attention in terms of citations in the research community, potentially because the research communities on IVA and IoT were focusing on other challenges at the time. However, commercial developments and products extended that early research demonstrated the importance for our daily life and thus caused a widespread interest by researchers in this convergence field over the last years.

7 The Convergence of AR and IoT

This convergence field brings together the field of AR, in which *virtual* content is overlaid or mixed with the *physical* world, and the field of IoT, in which *physical* devices are connected and *virtualized* for computer-controlled networked interaction.

7.1 Meta-Review

At the intersection of AR and IoT, we found 79 publications with 135 classifications in total. Figure 2 shows the publications by year, and Fig. 3 shows them by classification. Ten publications in this field had a citation count of four citations per year or higher. The average citation count in this field was 2.19 citations per year ($SD = 3.66$).

Papers in the field of AR ∩ IoT were mostly concerned with system development and theoretical frameworks (42), followed by application topics (22), evaluation (20), interaction (14), and survey papers (8). Twenty-six papers employed multimodal setups, and three papers developed collaborative designs.

As shown in Fig. 10, many of the publications in this field relied on mobile devices to implement and display their work (36), and fewer used HMDs (14), followed by flat screens (4), cross-displays (5), and projection-based AR (2). A good number of papers focusing on theoretical aspects did not employ a specific display for their work (18), and five papers developed frameworks that could be used on different types of displays.

As shown in Fig. 11 we see a high number of publications with light input/output and fewer with sound input/output which can be seen as human modalities or qualities of an IVA. Touch, motion, and temperature sensors were slightly more used in this field compared to others which also include specific IoT sensors in the environment affecting virtual content.

7.2 Impactful Papers and Trends

Most research at the intersection of AR and IoT considered AR as an interface for IoT devices that avoided limitations of traditional user interfaces in smart home

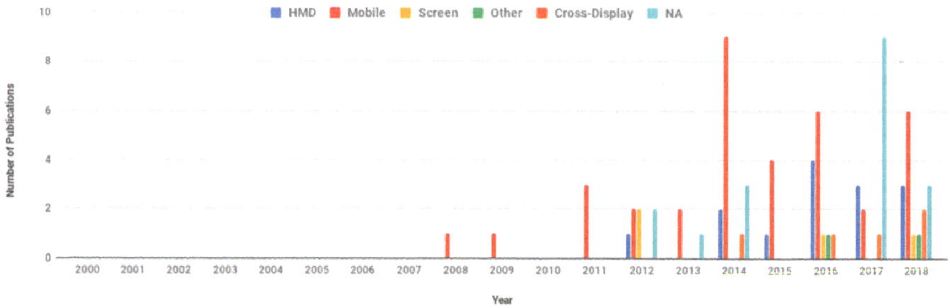

Fig. 10 Variety of displays used over time in the field of AR and IoT

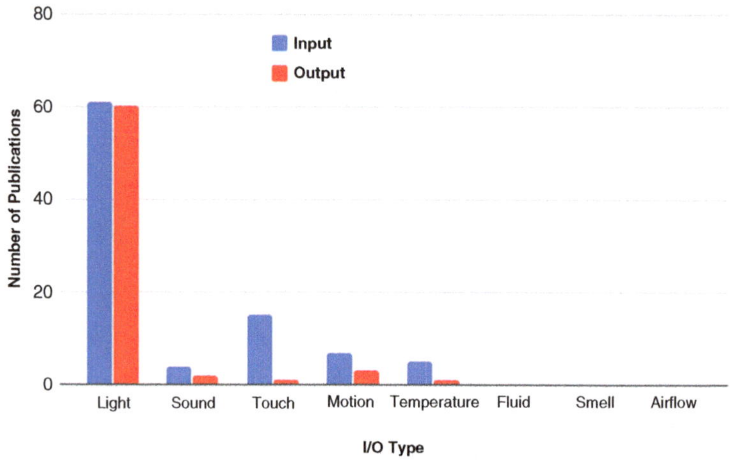

Fig. 11 Publications in the field of AR and IoT grouped based on the modalities used

environments. Gimenez and Pous [29] noted that "AR has recently been touted as one of the ideal interfaces for IoT." For instance, Garcia-Macias et al. [26] presented a mobile AR application to "browse" through the Internet of things. The rationale for using AR was that IoT unlike the traditional Internet constitutes a mix of virtual and physical entities. The authors developed a mobile AR application similar to a magic lens that lets users see smart objects in their environment when the mobile AR device is pointed at them, thus allowing the user to understand the virtual services provided by the IoT devices and interact with them. Wirtz et al. [76] presented a similar approach with the difference that the smart devices in the environment would not be directly connected to the Internet but connect to the mobile AR device on demand when pointing the AR device at the smart device. Another similar system was proposed by Heun et al. [34], called *The Reality Editor*, in which users hold a mobile AR device near a smart device to open an interface that allows users to reprogram the behavior of the smart device and to define relationships between smart devices in order to create new functionalities.

Kasahara et al. [38] further extended the approach with mobile AR as an interface for IoT to a novel class of actuated IoT devices. In their prototype, users could point a mobile AR device at an actuated smart connected appliance, and by touching the screen of the mobile AR device, users could then drag the target in the AR view, which the smart appliance would then match using an actuated motion platform.

Instead of using mobile AR devices, Lik-Hang and Pan [47] recently resumed different ways in which AR glasses and corresponding user interfaces could provide an accessible and intuitive user interface for interaction in a modern smart home environment. To note are in particular gestural and gaze-based interfaces, which allow users to control smart home appliances at a distance, e.g., by flicking a finger towards a light to turn it off without having to walk over to a light switch [43, 78]. Martin and LaViola [55] further introduced a *Transreality Interaction Platform (TrIP)* to "enable service ecosystems which situate virtual objects alongside virtualized physical objects and allow for novel ad-hoc interactions between humans, virtual, and physical objects in a transreality environment."

While smart home environments with AR interfaces were among the highest cited papers, they were not the only application fields for using AR and IoT. For example, Hao and Helo [32] emphasized on the importance of AR and IoT for the manufacturing industry. They investigated the application of AR and wearable technology to visualize information derived from IoT sensor networks in a cloud manufacturing environment.

Also, AR has been used as a test-bed for IoT environments. In particular, Seo et al. [69] designed multiple test-beds to experience and understand smart home environments and IoT networks in VR and AR with respect to context and awareness, before they are realized in an actual smart home environment.

In contrast to the aforementioned research that largely leveraged AR as a tool for interaction with IoT devices, it is interesting to note that researchers so far have not fully leveraged IoT devices as a tool to improve AR experiences.

In the research field of AR, various input and output devices have been prototyped, some have been connected to the Internet, and some have been commercially pursued. However, while many of these research prototypes developed in the field of AR could be realized in the form factor of an IoT device, this is hardly done so far. This might be explained by the perceived overhead involved in adhering to IoT protocols and standards and associated fears of added latency and suboptimal performance [68]. It may also be related to recent trends in the field of AR today that most developments are egocentric with respect to a physical device, i.e., sensors and displays are mounted on the user's body or head instead of them being placed throughout the environment, which makes it harder to think of them as Internet-connected *things*.

At the same time, we have to note that the steady integration of IoT capabilities into everyday devices all around us presents an opportunity for the addition of more allocentric pervasive real-world capabilities to AR. In particular, Jo and Kim [37] proposed an IoT extension to support scalable AR environments called *ARIoT*.

We predict that over the next years, further advances in AR user interfaces and technologies towards everyday use cases in a smart home environment will give rise to the standardized integration of IoT devices into AR frameworks and the development of new IoT devices specifically for the purpose of enhancing AR experiences with improved sensing and multimodal sensations, e.g., driven by the gaming industry.

8 The Nexus of IVA, IoT, and AR

While each of these convergent technology thrusts has shown benefits due to the integration of the research fields and can advance on their own, the nexus of the three concepts and technologies could enable further novel solutions and research directions. In particular, we see a vexing vista for the future of virtual agent technologies that becomes possible and tractable at the intersection of the three fields. This idea of a sophisticated futuristic agent could be manifested in the form of an embodied *augmented reality agent (ARA)* with pervasive *awareness*, *appearance*, and *abilities* through IVA intelligence and human-agent communication, IoT networks to sense and influence the real world, and AR display and sensor technology to seamlessly blend the agent with the real world. Such empowered agent technologies would be *transformational*.

Beyond one's physical "surroundings" [60] and the state of oneself and one's home, the contextual *awareness* would be *pervasive* in that it could extend to all matters of one's life that one allows (via privacy controls): one's health, calendar, daily fitness information, and even awareness of one's family members—shared awareness between the ARAs of the family homes. Indeed, the awareness could include ongoing assessments of one's physical and emotional health, which could be shared with a healthcare provider or family members. New privacy mechanisms would then be needed to control which aspects of awareness (and appearance and abilities) are shared.

The *appearance* of one's ARA would be *pervasive* in that it could move around within one's home, e.g., appearing anywhere in 3D if one is wearing an HMD, projected onto physical surroundings via *spatial augmented reality* [12, 62–65] in one's IoT-enabled television, in the display of one's IoT-enabled refrigerator, among others, to best match the current location of the human [8]. The ARA could be transferred to any network and device that one allows, e.g., one ARA could travel on one's mobile devices, one's car, and even one's office computer. One of many questions surrounding ARAs is whether there is value in seeing an embodied agent, even in cases where it is not necessary. For example, if one asks one's agent to check the calendar to see if one is free for a date tonight, and the agent just says "Yes," one might not trust the ARA's response as much as if it *appeared* to carry out some action that corresponds in some way to checking one's calendar. Among other things, we see the importance of new knowledge related to this "seeing is believing" and the importance of visual embodiments in general. More generally we see the need to advance knowledge on different agent representations based on their tasks and contexts and appearances that are adapted to each user [74], static or variable, which can range from single to multiple agent representations, e.g., when the agent is multitasking.

The *abilities* of one's ARA would be *pervasive* in that it could have *physical influence* throughout one's house under a variety of circumstances. For example, today one can purchase light switches and outlets that are compatible with various home automation protocols, e.g., Insteon, Z-Wave, and Zigbee. In the future, manufacturers could include AR features and data such as embedded textual and verbal descriptions, the ability to embed location information, etc. In terms of *social influence* [13], an ARA that is aware of one's dietary circumstances could coach one on what to eat using immediate caloric information from one's body-worn fitness devices, awareness of one's calendar for the day, etc. One's ARA could directly influence the person, e.g., by greeting one cheerfully and appreciatively when one returns home from work—the "Cheers effect" (American sitcom)—or could reach out to family members or friends if, for example, it becomes aware that one is overly tired or depressed. Among other directions, existing voice-controlled agents, home control, and custom technologies could be enhanced by giving ARAs the ability to control IoT devices.

There are two perspectives of ARA *affordances* [28] we propose should be examined: the affordances the ARA offers to the human and the affordances the environment offers to the ARA. For example, with respect to the former, embodied AR agents offer the opportunity to study the "seeing is believing" idea. With respect to the latter, ARA awareness of nonverbal communications by the human would support verbal instructions that are intentionally brief or ambiguous, such as "please turn of *that light*" while pointing to a particular lamp. This is in contrast to today's voice agents where one has to remember and use the precise name corresponding to each device, e.g., "please turn off the Living Room Floor Lamp!"

In summary, by extrapolating from the current developments discussed above, we believe that these affordances become tractable at the intersection of the three research fields AR, IVA, and IoT. As technology and prototypes are developed to address these challenges, we see much potential for an interesting and transformative research domain being established at the nexus of the three fields.

9 Conclusion

In this chapter, we resumed the body of literature at the convergence fields between augmented reality, intelligent virtual agents, and the Internet of things. We performed a literature search at the intersections between each two of these fields, and we analyzed the research topic categories of the publications and their impact as indicated by their citation count. We further highlighted and reviewed the most impactful papers and trends in these convergence fields. Our results show a large increase in publications in these convergence fields over the last years and much potential for future research. We predict that the three fields will grow in importance over the next years and attract interdisciplinary research teams to advance the technology development and improve our understanding of the challenges and goals in these fields. Finally, we presented our vision for the convergence of all three technologies.

References

1. Al-Fuqaha, A., Guizani, M., Mohammadi, M., Aledhari, M., & Ayyash, M. (2015). Internet of things: A survey on enabling technologies, protocols, and applications. *IEEE Communications Surveys & Tutorials, 17*(4), 2347–2376.
2. Al-Turjman, F. (2018). Information-centric framework for the Internet of Things (IoT): Traffic modeling & optimization. *Future Generation Computer Systems Journal, 80*, 63–75.
3. Al-Turjman, F., & Alturjman, S. (2018). Context-sensitive access in Industrial Internet of Things (IIoT) healthcare applications. *IEEE Transactions on Industrial Informatics, 14*, 2736–2744.
4. Anabuki, M., Kakuta, H., Yamamoto, H., & Tamura, H. (2000). *Welbo: An embodied conversational agent living in mixed reality space. CHI '00 Extended Abstracts on Human Factors in Computing Systems. Proceedings of the ACM SIGCHI Conference on Human Factors 659 in Computing Systems* (pp. 10–11).
5. Ashton, K. (2009). That 'internet of things' thing. *RFID journal, 22*(7), 97–114.
6. Austerjost, J., Porr, M., Riedel, N., Geier, D., Becker, T., Scheper, T., et al. (2018). Introducing a virtual assistant to the lab: A voice user Interface for the intuitive control of laboratory instruments. *SLAS TECHNOLOGY: Translating Life Sciences Innovation, 23*(5), 476–482.
7. Azuma, R. (1997). A survey of augmented reality. *Presence: Teleoperators & Virtual Environments, 6*(4), 355–385.
8. Barakonyi, I., & Schmalstieg, D. (2005). Augmented reality agents in the development pipeline of computer entertainment. *International Conference on Entertainment Computing* (pp. 345–356).
9. Barakonyi, I., & Schmalstieg, D. (2006). Ubiquitous animated agents for augmented reality. *2006 IEEE/ACM International Symposium on Mixed and Augmented Reality* (pp. 145–154).
10. Barakonyi, I., Psik, T., & Schmalstieg, D. (2004). Agents that talk and hit back: Animated agents in augmented reality. *IEEE and ACM International Symposium on Mixed and Augmented Reality* (pp. 141–150).
11. Barakonyi, I., Weilguny, M., Psik, T., & Schmalstieg, D. (2005). Monkey Bridge: Autonomous agents in augmented reality games. *Proceedings of the 2005 ACM SIGCHI International Conference on Advances in Computer Entertainment Technology* (pp. 172–175).
12. Bimber, O., & Raskar, R. (2005). *Spatial augmented reality: Merging real and virtual worlds*. Wellesley, MA: A.K. Peters.
13. Blascovich, J. (2002). Social influence within immersive virtual environments. In *The social life of avatars* (pp. 127–145). London: Springer.
14. Blum, L., Wetzel, R., McCall, R., Oppermann, L., & Broll, W. (2012). The final TimeWarp: Using form and content to support player experience and presence when designing location-aware mobile augmented reality games. *Proceedings of the Designing Interactive Systems Conference on* (pp. 711–720).
15. Bölöni, L., & Turgut, D. (2017). Value of information based scheduling of cloud computing resources. *Future Generation Computer Systems Journal, 71*, 212–220.

16. Campagna, G., Ramesh, R., Xu, S., Fischer, M., & Lam, M. (2017). Almond: The architecture of an open, crowdsourced, privacy-preserving, programmable virtual assistant. *WWW '17 Proceedings of the 26th International Conference on World Wide Web* (pp. 341–350).
17. Charles, F., Cavazza, M., Mead, S., Martin, O., Nandi, A., & Marichal, X. (2004). Compelling experiences in mixed reality interactive storytelling. *Proceedings of the 2004 ACM SIGCHI International Conference on Advances in Computer Entertainment Technology* (pp. 32–40).
18. Chekhlov, D., Gee, A., Calway, A., & Mayol-Cuevas, W. (2007). Ninja on a plane: Automatic discovery of physical planes for augmented reality using visual SLAM. *6th IEEE and ACM International Symposium on Mixed and Augmented reality* (pp. 153–156).
19. Chung, H., Iorga, M., Voas, J., & Lee, S. (2017). Alexa, Can I Trust You? *IEEE Computer, 50*(9), 100–104.
20. Chung, H., Park, J., & Lee, S. (2017). Digital forensic approaches for Amazon Alexa ecosystem. *Digital Investigation, 22*, S15–S25.
21. Daher, S., Kim, K., Lee, M., Bruder, G., Schubert, R., Bailenson, J., et al. (2017). Can social presence be contagious? Effects of social presence priming on interaction with virtual humans. *2017 IEEE Symposium on 3D User Interfaces (3DUI)* (pp. 201–202).
22. Dey, A., Billinghurst, M., Lindeman, R., & Swan, J. (2018). A systematic review of 10 years of augmented reality usability studies: 2005 to 2014. *Frontiers in Robotics and AI, 5*, 37.
23. Dow, S., Mehta, M., Harmon, E., MacIntyre, B., & Mateas, M. (2007). Presence and engagement in an interactive drama. *Proceedings of the SIGCHI Conference on Human Factors in Computing Systems* (pp. 1475–1484).
24. Druga, S., Williams, R., Breazeal, C., & Resnick, M. (2017). "Hey Google is it OK if I eat you?": Initial explorations in child-agent interaction. *Proceedings of the 2017 Conference on Interaction Design and Children* (pp. 595–600).
25. Evans, D. (2011). The internet of things: How the next evolution of the internet is changing everything. *Cisco Blog, 1*, 1–1.
26. García-Macías, J. A., Alvarez-Lozano, J., Estrada-Martinez, P., & Avilés-López, E. (2011). Browsing the internet of things with sentient visors. *Computer, 44*(5), 46–52.
27. Gatebox, Inc. (2018). Retrieved from https://gatebox.ai.
28. Gibson, J. (1979). *The ecological approach to visual perception*. Dallas: Houghtom Mifflin.
29. Gimenez, R., & Pous, M. (2010). Augmented reality as an enabling factor for the internet of things. *Proceedings of the W3C Workshop: Augmented Reality on the Web*.
30. Growing Convergence Research. (2018). Retrieved from National Science Foundation: https://www.nsf.gov/news/special_reports/big_ideas/convergent.jsp.
31. Hantono, B., Nugroho, L., & Santosa, P. (2016). Review of augmented reality agent in education. *2016 6th International Annual Engineering Seminar (InAES)* (pp. 150–153).
32. Hao, Y., & Helo, P. (2017). The role of wearable devices in meeting the needs of cloud manufacturing: A case study. *Robotics and Computer-Integrated Manufacturing, 45*, 168–179.
33. Helal, A., Cho, K., Lee, W., Sung, Y., Lee, J., & Kim, E. (2012). 3D modeling and simulation of human activities in smart spaces. *2012 9th International Conference on Ubiquitous Intelligence and Computing and 9th International Conference on Autonomic and Trusted Computing* (pp. 112–119).
34. Heun, V., Hobin, J., & Maes, P. (2013). Reality editor: Programming smarter objects. *Proceedings of the 2013 ACM conference on Pervasive and Ubiquitous Computing Adjunct Publication* (pp. 307–310).
35. Hoffman, D., & Novak, T. (2018). Consumer and object experience in the internet of things: An assemblage theory approach. *Journal of Consumer Research, 44*(6), 1178–1204.
36. Holz, T., Campbell, A., O'Hare, G., Stafford, J., Martin, A., & Dragone, M. (2011). MiRA-Mixed Reality Agents. *International Journal of Human-Computer Studies/International Journal of Man-Machine Studies, 69*(4), 251–268.
37. Jo, D., & Kim, G. (2016). ARIoT: Scalable augmented reality framework for interacting with internet of things appliances everywhere. *IEEE Transactions on Consumer Electronics, 62*(3), 334–340.

38. Kasahara, S., Niiyama, R., Heun, V., & Ishii, H. (2013). exTouch: Spatially-aware embodied manipulation of actuated objects mediated by augmented reality. *Proceedings of the 7th International Conference on Tangible, Embedded and Embodied Interaction* (pp. 223–228).
39. Kim, K., Bruder, G., Maloney, D., & Welch, G. (2016). The influence of real human personality on social presence with a virtual human in augmented reality. *ICAT-EGVE '16 proceedings of the 26th International Conference on Artificial Reality and Telexistence and the 21st Eurographics Symposium on Virtual Environments* (pp. 115–122).
40. Kim, K., Bruder, G., & Welch, G. (2017). Exploring the effects of observed physicality conflicts on real-virtual human interaction in augmented reality. *Proceedings of the 23rd ACM Symposium on Virtual Reality Software and Technology* (p. 31).
41. Kim, K., Billinghurst, M., Bruder, G., Duh, H. B.-L., & Welch, G. (2018). Revisiting trends in augmented reality research: A review of the 2nd decade of ISMAR (2008–2017). *IEEE Transactions on Visualization and Computer Graphics (TVCG) Special Issue on the International Symposium on Mixed and Augmented Reality (ISMAR).*
42. Knote, R., Janson, A., Eigenbrod, L., & Söllner, M. (2018). The what and how of smart personal assistants: Principles and application domains for IS research. In: Multikonferenz Wirtschaftsinformatik (MKWI). Lüneburg: Germany.
43. Kollee, B., Kratz, S., & Dunnigan, A. (2014). Exploring gestural interaction in smart spaces using head mounted devices with ego-centric sensing. *Proceedings of the 2nd ACM symposium on spatial user interaction* (pp. 40–49).
44. Kotranza, A., & Lok, B. (2008). Virtual human + tangible interface = mixed reality human an initial exploration with a virtual breast exam patient. *2008 IEEE Virtual Reality Conference* (pp. 99–106).
45. Kotranza, A., Lok, B., Deladisma, A., Pugh, C., & Lind, D. (2009). Mixed reality humans: Evaluating behavior, usability, and acceptability. *IEEE Transactions on Visualization and Computer Graphics, 15*(3), 369–382.
46. Krum, D., Suma, E., & Bolas, M. (2012). Augmented reality using personal projection and retroreflection. *Ubiquitous Computing, 16*(1), 17–26.
47. Lee, L.H. & Hui, P. (2018). Interaction Methods for Smart Glasses: A survey. *IEEE Access*, (pp. 28712–28732).
48. Lee, M., Kim, K., Daher, S., Raij, A., Schubert, R., Bailenson, J., et al. (2016a). The wobbly table: Increased social presence via subtle incidental movement of a real-virtual table. *2016 IEEE Virtual Reality (VR)* (pp. 11–17).
49. Lee, W., Cho, S., Chu, P., Vu, H., Helal, S., Song, W., et al. (2016b). Automatic agent generation for IoT-based smart house simulator. *Neurocomputing, 209*, 14–24.
50. Lee, M., Bruder, G., & Welch, G. (2017). Exploring the effect of vibrotactile feedback through the floor on social presence in an immersive virtual environment. *2017 IEEE Virtual Reality (VR)* (pp. 105–111).
51. Lee, M., Bruder, G., Hollerer, T., & Welch, G. (2018). Effects of unaugmented periphery and vibrotactile feedback on proxemics with virtual humans in AR. *IEEE Transactions on Visualization and Computer Graphics, 24*(4), 1525–1534.
52. Lok, B., Chuah, J., Robb, A., Cordar, A., Lampotang, S., Wendling, A., et al. (2014). Mixed-reality humans for team training. *IEEE Computer Graphics and Applications, 34*(3), 72–75.
53. López, G., Quesada, L., & Guerrero, L. A. (2017). Alexa vs. Siri vs. Cortana vs. Google assistant: A comparison of speech-based natural user interfaces. *Proceedings of the International Conference on Applied Human Factors and Ergonomics* (pp. 241–250).
54. Magic Leap, Inc. (2018). Retrieved from https://www.magicleap.com.
55. Martin, K., & Laviola, J. (2016). The transreality interaction platform: Enabling interaction across physical and virtual reality. *2016 IEEE International Conference on Internet of Things (iThings) and IEEE Green Computing and Communications (GreenCom) and IEEE Cyber, Physical and Social Computing (CPSCom) and IEEE Smart Data (SmartData)* (pp. 177–186).
56. Mayle, A., Bidoki, N. H., Masnadi, S., Bölöni, L., & Turgut, D. (2017). Investigating the value of privacy within the internet of things. *Proceedings of IEEE GLOBECOM* (pp. 1–6).

57. Milgram, P., Takemura, H., Utsumi, A., & Kishino, F. (1995). Augmented reality: A class of displays on the reality-virtuality continuum. *Telemanipulator and Telepresence Technologies, 2351*, 282–292.
58. Newsroom Gartner. (2016). Retrieved from Gartner says worldwide spending on VPA-enabled wireless speakers will top $2 billion by 2020: https://www.gartner.com/newsroom/id/3464317.
59. Norouzi, N., Kim, K., Hochreiter, J., Lee, M., Daher, S., Bruder, G., et al. (2018). A systematic survey of 15 years of user studies Published in the Intelligent virtual agents conference. *International Conference on Intelligent Virtual Agents (IVA)*.
60. Papagiannis, H. (2017). *Augmented human: How technology is shaping the new reality*. Bejing: O'Reilly Media.
61. Paul, Z., Margarita, P., Vasilis, M., & George, P. (2016). Life-sized group and crowd simulation in Mobile AR. *Proceedings of the 29th International Conference on Computer Animation and Social Agents* (pp. 79–82).
62. Raskar, R. (2001). *Projector-based three dimensional graphics*. Chapel Hill: University of North Carolina.
63. Raskar, R., Welch, G., Cutts, M., Lake, A., Stesin, L., & Fuchs, H. (1998). The office of the future: A unified approach to image-based modeling and spatially immersive displays. *Proceedings of the 25th Annual Conference on Computer Graphics and Interactive Techniques* (pp. 179–188).
64. Raskar, R., Welch, G., & Chen, W.-C. (1999). Table-top spatially-augmented realty: Bringing physical models to life with projected imagery. *Proceedings 2nd IEEE and ACM International Workshop on Augmented Reality (IWAR'99)* (pp. 64–71).
65. Raskar, R., Welch, G., Low, K.-L., & Bandyopadhyay, D. (2001). Shader lamps: Animating real objects with image-based illumination. *Proceedings of the 12th Eurographics Workshop on Rendering Techniques* (pp. 89–102).
66. Reis, A., Paulino, D., Paredes, H., & Barroso, J. (2017). Using intelligent personal assistants to strengthen the Elderlies' social bonds. *International Conference on Universal Access in Human-Computer Interaction* (pp. 593–602).
67. Robb, A., Cordar, A., Lampotang, S., White, C., Wendling, A., & Lok, B. (2015). Teaming up with virtual humans: How other people change our perceptions of and behavior with virtual teammates. *IEEE Transactions on Visualization and Computer Graphics, 21*(4), 511–519.
68. Salman, T., & Jain, R. (2017). A survey of protocols and standards for internet of things. *Advanced Computing and Communications, 1*(1), 1–20.
69. Seo, D., Kim, H., Kim, J., & Lee, J. (2016). Hybrid reality-based user experience and evaluation of a context-aware smart home. *Computers in Industry, 76*, 11–23.
70. Skarbez, R., Welch, G., Brooks, F., & Whitton, M. (2017). Coherence changes gaze behavior in virtual human interactions. *2017 IEEE Virtual Reality (VR)* (pp. 287–288).
71. Soda, S., Nakamura, M., Matsumoto, S., Izumi, S., Kawaguchi, H., & Yoshimoto, M. (2012). Implementing virtual agent as an interface for smart home voice control. *2012 19th Asia-Pacific Software Engineering Conference*, 1, pp. 342–345.
72. Sutherland, I. (1968). A head-mounted three dimensional display. *Proceedings of the December 9–11, 1968, Fall Joint Computer Conference, Part I on* (pp. 757–764).
73. Turgut, D., & Bölöni, L. (2017, September). Value of information and cost of privacy in the internet of things. *IEEE Communications Magazine, 55*(9), 62–66.
74. Vugt, H., Bailenson, J., Hoorn, J., & Konijn, E. (2010). Effects of facial similarity on user responses to embodied agents. *ACM Transactions on Computer-Human Interaction, 17*(2), 7.
75. Wagner, D., Billinghurst, M., & Schmalstieg, D. (2006). How real should virtual characters be. *Proceedings of the 2006 ACM SIGCHI International Conference on Advances in Computer Entertainment Technology* (p. 57).
76. Wirtz, H., Rüth, J., Serror, M., Link, J., & Wehrle, K. (2014). Opportunistic interaction in the challenged internet of things. *Proceedings of the 9th ACM MobiCom Workshop on Challenged Networks* (pp. 7–12).

77. Zehtabian, S., Khodadadeh, S., Pearlman, R., Willenberg, B., Kim, B., Turgut, D., et al. (2018). Supporting rehabilitation prescription compliance with an IoT-augmented four-legged walker. *Workshop on AI for Aging, Rehabilitation and Independent Assisted Living (ARIAL'18) in Conjunction with International Joint Conference on Artificial Intelligence (IJCA'18).*
78. Zhang, B., Chen, Y.-H., Tuna, C., Dave, A., Li, Y., Lee, E., et al. (2014). HOBS: Head orientation-based selection in physical spaces. *Proceedings of the 2nd ACM Symposium on Spatial User Interaction* (pp. 17–25).
79. Cisco. (2018). Retrieved from Internet of things at a glance: https://www.cisco.com/c/dam/en/us/products/collateral/se/internet-of-things/at-a-glance-c45-731471.pdf.

Improving the Physical Layer Security of IoT-5G Systems

Jehad M. Hamamreh

1 Introduction

Security is becoming not only a necessary and critical requirement in communication networks but also a powerful mean and tool to maintain the sustainability and usefulness of the digital world we are living in [1, 2]. It has recently become very obvious that the super benefits in terms of new Internet of Things (IoT)-based services and applications that future 5G and beyond wireless networks are expected to provide cannot be practically realized without ensuring security and having it as a high priority [3, 4]. Particularly, security forms a formidable challenge for the new emerging massive and critical IoT-5G type of services such as massive machine-type communication (mMTC) and ultra-reliable low-latency communication (URLLC) due to their unique features and requirements [5, 6].

Among the many important security aspects and goals including confidentiality, privacy, authentication, integrity, availability, and so forth, confidentiality comes at the highest priority. Providing confidentiality is a challenging task to achieve in wireless systems mainly because of the broadcast nature of the wireless transmission, where the transmitted signals can be captured and recorded by malicious users in the networks in order to intercept the undergoing communication between the legitimate parties. This results in not only eavesdropping their data but also in using this eavesdropped data to launch more attacks such as identity-based attack, denial of service (DoS), man-in-the-middle, data modification, session hijacking, spoofing (impersonation), and sniffing. These critical vulnerabilities could significantly affect the authenticity, confidentiality, integrity, and availability of the communication link

J. M. Hamamreh (✉)
Department of Electrical-Electronics Engineering, Antalya International (Bilim) University, Antalya, Turkey
e-mail: jehad.hamamreh@antalya.edu.tr

F. Al-Turjman (ed.), *Artificial Intelligence in IoT*,
Transactions on Computational Science and Computational Intelligence,
https://doi.org/10.1007/978-3-030-04110-6_2

between legitimate users [7, 8]. This necessitates providing security defense against eavesdropping, which is considered to be one of the most critical attacks that is needed as a prerequisite to launch any kind of attacks in the network.

Traditionally, confidentiality against eavesdropping has been tackled using cryptography and encryption-based approaches. However, the tremendous technological advances in future wireless networks (e.g., 5G and beyond networks) substantiate the need for new alternative security methods that can meet and address new challenges that did not exist before. This makes the classically used cryptographic techniques unsuitable, ineffective, and even impractical in future wireless networks and services due to many hindrances, among which, we mention the key distribution and management processes for the legitimate parties. These are extremely difficult tasks, especially in large-scale, dense, and heterogeneous wireless networks, where a massive amount of smart IoT devices are simultaneously connected to the network, causing excessive complexity, high overhead, and costly computational processes. Besides, the management and control frames exchanged between communication entities are usually not protected. More importantly, the emergence of new wireless technologies like IoT-based mMTC, 5G-Tactile Internet, vehicular communication for autonomous driving, remote surgery, instant control for sensitive IoT actuators, etc. makes current encryption-based methods unsuitable since these kind of technologies are naturally delay-sensitive, power-limited, and processing-restricted.

To address this, physical layer security based on keyless approach has emerged as a new concept and strong alternative that can supplement and may even replace encryption-based approaches [9–11]. The basic idea of physical layer security is to exploit channel characteristics such as noise, fading, dispersion, and interference, in order to provide secrecy against eavesdropping [9, 12].

Due to the fact that orthogonal frequency-division multiplexing (OFDM) is the most commonly used waveform in currently existing systems and is expected to keep its dominance with different numerologies in future wireless systems like 5G, securing OFDM waveform has drawn the attention of many researchers. It is worth mentioning that besides developing techniques tailored to common transmit waveforms like OFDM, there have recently been some efforts to design new inherently secure waveforms as in [13–15].

In the literature, several OFDM-based security techniques have been proposed. These techniques can be categorized from a high-level viewpoint into four main enabling schemes. First are secret key-based schemes, in which secret random sequences are generated from the channel and then used to encrypt the transmitted data on either the application layer [16] or the physical layer such as dynamic coordinate interleaving and constellation rotation schemes [17]. Second are adaptive transmission-based schemes, in which the transmission parameters are adjusted to just meet the quality of service (QoS) requirements of only the legitimate receiver [18]. Among these techniques are optimal power allocation [19], adaptive modulation with hybrid automatic repeat request (HARQ) [20], adaptive precoding and interleaving [21], fading-based subcarrier deactivation schemes [22], and channel shortening [23]. Third are artificial noise (AN)-based schemes [24], in which AN is designed based on the legitimate receiver's channel so that it only harms the eavesdropper's reception while maintaining an interference-free reception at the legitimate user. Fourth are schemes that can exploit OFDM transceiver impairments [25] or conceal some key features in the OFDM signal to provide secrecy [26].

Among the aforementioned techniques, OFDM-SIS (subcarrier index selection) [27] is a recently proposed flexible and adaptive technique that not only provides secrecy but also improves the reliability performance with minimal complexity, making it a very attractive technique for future secure IoT-5G systems. However, although the optimal subcarrier index selection technique (OFDM-SIS) can provide a relatively good secrecy performance, its secrecy gap is limited and may not be sufficient in specific scenarios, where it is highly desirable to increase and ensure secrecy even when Eve's average signal-to-noise ratio (SNR) is much higher than that of Bob.

Therefore, this work comes to address the aforementioned limitation of OFDM-SIS by enhancing its secrecy further and maintaining a good secrecy gap over all SNR ranges. In the proposed scheme, named as OFDM-SIS with artificially interfering signals (OFDM-SIS-AIS), the whole OFDM block is divided into subblocks, each containing good and bad subcarriers. The good subcarriers, corresponding to high subchannel gains, are used for data transmission in order to maximize the SNR at only the legitimate receiver, while the rest (the bad ones) are injected with AIS that can only degrade Eve's reception while not affecting Bob's performance whatsoever. The provided results prove the effectiveness of the proposed design in enhancing the secrecy gap considerably over all SNR regimes while keeping the bit error rate (BER) performance of the legitimate receiver as that achieved by OFDM-SIS.

The rest of this chapter is organized as follows. The system model and its preliminaries are described in Sect. 2. The details of the proposed OFDM-SIS-AIS scheme are revealed in Sect. 3. Computer simulation results are exhibited and discussed in Sect. 4. Finally, a concise conclusion is drawn in Sect. 5.

Notations: Vectors are denoted by bold small letters, whereas matrices are denoted by bold capital letters. $\mathbf{I}$ is the $N \times N$ identity matrix. The transpose, Hermitian, and inverse of a matrix are symbolized by $(\cdot)^{\mathrm{T}}$, $(\cdot)^{\mathrm{H}}$ and $(\cdot)^{-1}$, respectively.

2 Preliminaries and System Model

A single-input single-output (SISO) OFDM system is adopted, where a transmitting node (Alice) tries to communicate confidentially with a legitimate receiving node (Bob), whereas an eavesdropping node (Eve) attempts to intercept the communication link between the two legitimate nodes (Alice and Bob) as shown in Fig. 1. The channels of both Bob and Eve, denoted by $\mathbf{h_b} \in \mathbb{C}^{[1\times L]}$ and $\mathbf{h_e} \in \mathbb{C}^{[1\times L]}$, respectively, are assumed to experience multipath slowly varying channels with L exponentially decaying taps, each with Rayleigh fading distribution. Moreover, as

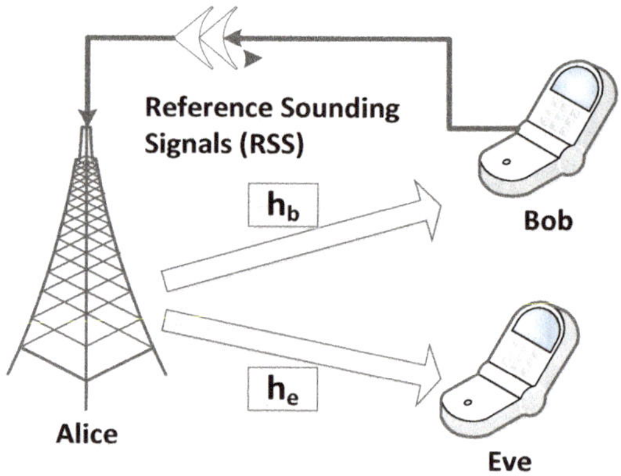

Fig. 1 A simplified system model of the considered PHY security scenario

Eve is a passive node in practice, Alice is assumed to have no knowledge of Eve's channel.

Moreover, both Bob and Eve are assumed to experience uncorrelated channels as the channel varies according to the positions of communicating nodes [14]. In addition, time division duplexing (TDD) system is adopted, where Alice and Bob estimate the channel state information (CSI) of their common link by sending sounding reference signals (SRS). This will prevent Eve from accessing or having the knowledge of the CSI of the legitimate link [14] as there is no explicit CSI feedback.

At Alice, N number of frequency-domain complex data symbols is transmitted. Thus, the frequency-domain OFDM symbol can be represented as $\mathbf{s} = \left[s_0\ s_1\ \ldots\ s_{N-1}\right]^{\mathrm{T}} \in \mathbb{C}^{[N\times 1]}$. The OFDM symbol is then interleaved in order to eliminate the correlation between the subchannels and make the effective channel response look random and completely independent as shown in Fig. 2. This is performed so that we can ensure distributing the deep-faded subchannels uniformly over the whole OFDM symbol and thus guarantee to experience a few deep-faded subchannels in each subblock. In this work, we consider using an adaptive CSI-dependent interleaver, denoted by a unitary matrix $\mathbf{R}$ of size $N \times N$, where the entries of each column are all zeros except a single entry of value equals to one at the position of the subcarrier to be permutated [21]. We select CSI-based adaptive interleaving as it is known to be the best in terms of mitigating burst errors (or consecutive deep-faded subchannels) and make them uniformly distributed over the whole OFDM block when the CSI is available at Tx [28].

It is worth mentioning that the interleaver design devised in [21] was perceived as a kind of precoder due to the fact that $\mathbf{R}$ was extracted by applying singular value decomposition (SVD) on the diagonal matrix of the channel amplitude frequency response and then taking the right unitary matrix, resulting from the decomposition, as the interleaver. For more details on this, we refer the readers to [21, 28]. It should also be mentioned that we do not consider in our performance evaluation the secrecy level that can be obtained by the channel-based adaptive interleaver design as it has

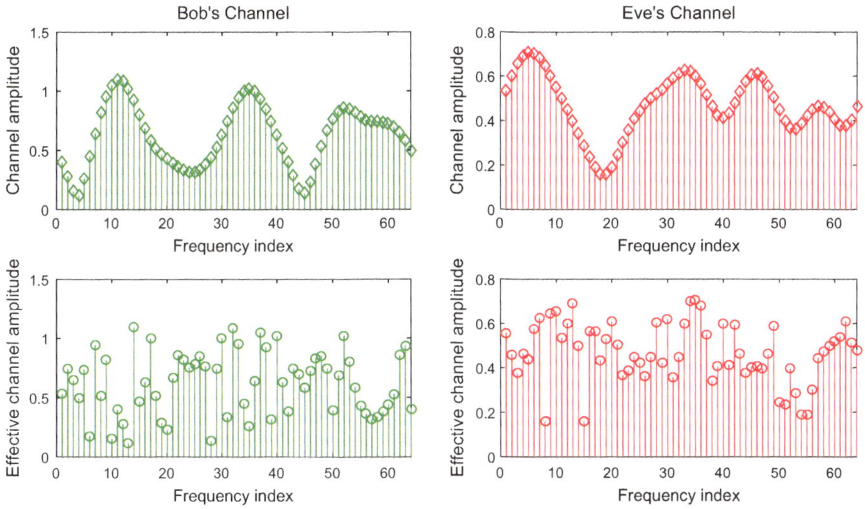

Fig. 2 Bob's and Eve's channel frequency responses alongside their effective channels after interleaving, i.e., $\mathbf{H}_\mathbf{b}^\mathbf{f}\mathbf{R}$ and $\mathbf{H}_\mathbf{e}^\mathbf{f}\mathbf{R}$ (shown in the lower part of the figure)

already been investigated in the literature [21]. Rather, we focus on the secrecy level obtained by the optimal subcarrier index selection along with artificially interfering signals injection.

Similar to OFDM-IM [29], where the whole OFDM block is split into subblocks and only a subset of the available subcarriers in each subblock is utilized for data transmission, here we adopt a similar structure; however, in our technique, the subcarriers are selected based on the channel of the legitimate receiver to improve secrecy. This is different from OFDM-IM, where the subcarriers are selected based on the incoming data to convey information by the subcarrier indices so that better reliability can be achieved at the expense of a minor spectral efficiency loss. The details of the proposed OFDM-SIS-AIS security technique will be explained in the next section.

The interleaved block ($\mathbf{Rs}$) is then passed through an IFFT process $\mathbf{F}^\mathrm{H} \in \mathbb{C}^{[N\times N]}$, which basically maps the data points to orthogonal subcarriers, where $\mathbf{F}$ is the discrete Fourier transform matrix. To preclude the inter-symbol interference, a cyclic prefix (CP) of length L is inserted by using the CP appending matrix $\mathbf{C} \in \mathbb{R}^{[(N+L)\times N]}$. Thus, the transmitted baseband signal by Alice (before considering the proposed design) can be represented as

$$\mathbf{x} = \mathbf{C}\mathbf{F}^\mathrm{H}\mathbf{Rs} \in \mathbb{C}^{[(N+L)\times 1]}. \tag{1}$$

After the signal $\mathbf{x}$ passes through the channel and reaches both Bob and Eve, each one of them will first discard the CP part of the signal using the matrix $\mathbf{D} \in \mathbb{R}^{[N\times(N+L)]}$ and then perform an FFT process using the matrix $\mathbf{F} \in \mathbb{C}^{[N\times N]}$ to transform the signal into the frequency domain. Thus, the net-received signal vector with dimensions $N \times 1$ at Bob after performing the aforementioned operations can be given in a linear matrix representation form as follows:

$$\mathbf{y_b} = \mathbf{FD}\left(\mathbf{H_b}\mathbf{C}\mathbf{F}^{\mathrm{H}}\mathbf{R}\mathbf{s} + \mathbf{z_b}\right) \tag{2}$$

$$= \mathbf{H}^{\mathbf{f}}_{\mathbf{b}}\mathbf{R}\mathbf{s} + \hat{\mathbf{z}}_{\mathbf{b}}. \tag{3}$$

On the other hand, at Eve, the captured signal after the FFT process can be formulated as

$$\mathbf{y_e} = \mathbf{FD}\left(\mathbf{H_e}\mathbf{C}\mathbf{F}^{\mathrm{H}}\mathbf{R}\mathbf{s} + \mathbf{z_e}\right) \tag{4}$$

$$= \mathbf{H}^{\mathbf{f}}_{\mathbf{e}}\mathbf{R}\mathbf{s} + \hat{\mathbf{z}}_{\mathbf{e}}. \tag{5}$$

In this model, $\mathbf{H_b} \in \mathbb{C}^{[(N+L)\times(N+L)]}$ and $\mathbf{H_e} \in \mathbb{C}^{[(N+L)\times(N+L)]}$ are the Toeplitz matrices corresponding to the channel impulse responses of both Bob and Eve, whereas $\mathbf{H}^{\mathbf{f}}_{\mathbf{b}} = \mathbf{FDH_bCF}^{\mathrm{H}} = \mathrm{diag}[H_{b_1}, \cdots, H_{b_N}] \in \mathbb{C}^{[N\times N]}$ and $\mathbf{H}^{\mathbf{f}}_{\mathbf{e}} = \mathbf{FDH_eCF}^{\mathrm{H}} = \mathrm{diag}[H_{e_1}, \cdots, H_{e_N}] \in \mathbb{C}^{[N\times N]}$ are the diagonal matrices corresponding to the channel frequency responses of Bob and Eve, respectively. Note that H_{b_i} and H_{e_i} for $1 \leq i \leq N$ denote the subchannel frequency response of the ith subcarrier with respect to Bob and Eve, respectively. The vectors $\mathbf{z_b}$ and $\mathbf{z_e}$ are formed by the samples of the zero-mean complex additive white Gaussian noise (AWGN) with variances of σ_b^2 and σ_e^2 at Bob and Eve, respectively, while $\hat{\mathbf{z}}_{\mathbf{b}}$ and $\hat{\mathbf{z}}_{\mathbf{e}}$ are the Fourier-transformed versions of the noise vectors at Bob and Eve, respectively.

3 Review of OFDM-Subcarrier Index Selection (OFDM-SIS)

In the OFDM-SIS scheme [27], the transmitted OFDM block, i.e., $\mathbf{s}$, is first divided and partitioned into a set of smaller subblocks, each containing K number of subcarriers. The basic idea of the proposed scheme is to enlarge the gap between Bob and Eve's capacities by making the effective SNR at Bob higher than that at Eve for a given channel frequency response. This is achieved by dividing the K subcarriers in each subblock into two subsets: the first, which includes the subcarriers corresponding to the highest subchannel gains, is used for data transmission (to maximize the effective SNR at Bob), whereas the second, which includes the subcarriers corresponding to the lowest channel gains, is suppressed and nulled (i.e., not used for data transmission).

Particularly, for each subblock, a set of M out of K subcarriers is optimally selected to maximize the SNR at Bob, while the remaining $K-M$, which are not used for data transmission, are filled with zeros. Here, $\zeta = M/K$ is defined as the subcarrier activation ratio of the number of selected subcarriers to the number of available subcarriers in each subblock. Note that the SNR of Bob over each subcarrier can be given by $SNR_{b_i} = \gamma_b = \frac{P|H_{b_i}|^2}{\sigma_b^2}$, where P is the power allocated to each subcarrier. Now, the problem of the optimal selection of the indices of M subcarriers corresponds to solving the below optimization problem for all possible subcarrier combinations, given as

$$\{c_1^{opt}, \cdots, c_M^{opt}\} = \arg \max_{\{c_1, \cdots, c_M\} \in \mathcal{A}_M} SNR_{b_{[c_1, \cdots, c_M]}}, \tag{6}$$

where $\mathcal{A}_M$ denotes the set of all possible subcarrier combinations with M selected out of K subcarriers and $SNR_{b_{[c_1, \cdots, c_M]}}$ is the sum of SNRs of the M selected subcarriers in each subblock. Since uniform power allocation is used for all subcarriers, the aforementioned problem boils down to selecting the subcarriers corresponding to the best subchannel gains.

This can be given as below

$$\{c_1^{opt}, c_2^{opt}, \cdots, c_M^{opt}\} = \arg \max_{\{c_1, \cdots, c_M\} \in \mathcal{A}_M} H_{b_{[c_1, \cdots, c_M]}}. \tag{7}$$

That is, finding all possible subcarrier combinations, i.e., $\binom{K}{M} = \frac{K!}{M!(K-M)!}$, may cause huge complexity, especially when the block size is very large. Therefore, it is important to considerably reduce the complexity of solving the above problem. This is possible when the whole OFDM block is divided into smaller parts to decrease the size of the search space. Moreover, it is also required to guarantee that, in each subblock, the subcarriers have to experience independent and different high and low subchannel gains so that the high ones with respect to Bob can be used for data transmission, while the low ones are suppressed (i.e., filled with zeros).

Now, to further minimize the complexity of the optimization problem, Alice can select M $(1 \leq M \leq K)$ out of K subcarriers that maximizes the effective instantaneous SNR at Bob in each subblock by first ranking the subcarriers based on their instantaneous channel gains in a descending order, i.e., $\{\|H_{b1}\|^2 \geq \|H_{b2}\|^2 \geq \cdots \geq \|H_{bK}\|^2\}$. Then, Alice selects the first M indices of the subcarriers corresponding to the sorted subchannel gains.

In this scheme, the transmitted baseband signal by Alice can be reformulated as

$$\mathbf{x} = \mathbf{C}\mathbf{F}^{\mathbf{H}}\mathbf{R}\mathbf{P}\begin{bmatrix}\mathbf{s}_d \\ \mathbf{s}_z\end{bmatrix} \in \mathbb{C}^{[(N+L)\times 1]}, \tag{8}$$

where $\mathbf{s}_d = [s_{(1)}, s_{(2)}, \ldots, s_{(N-N_r)}]^{\mathrm{T}}$ is the vector of $N_d = N - N_r = \zeta N$ frequency data symbols and $\mathbf{s}_z = [s_{(1)}, s_{(2)}, \ldots, s_{(N_r)}]^T$ is the zero vector of $N_r = (1-\zeta)N$ frequency points (i.e., nulled subcarriers). $\mathbf{P}$ is the permutation matrix, which determines the positions of the data and the nulled subcarriers within the subblocks of each OFDM symbol.

4 Proposed OFDM-Subcarrier Index Selection with Artificially Interfering Signals (OFDM-SIS-AIS)

Although the OFDM-SIS technique discussed in the previous section can provide a relatively good secrecy performance as demonstrated in [27], its secrecy gap may not be sufficient enough in specific scenarios, where it is highly desirable to increase the secrecy gap even when Eve's average SNR is higher than that of Bob. In addition, despite the fact that OFDM-SIS technique is motivated (besides its low complexity and better reliability) by its capability to work in the worst security scenario, where the CSI of the legitimate link can be accessed by Eve due to using explicit feedback as is the case in FDD systems, the technique adheres and maintains its applicability in TDD systems as well and can be modified in such a way to boost the achievable secrecy performance (especially in scenarios where Eve's SNR is higher than that of Bob). This is possible in TDD systems by utilizing the fact that Eve has no knowledge on the CSI between the legitimate parties due to using channel reciprocity-dependent sounding techniques for CSI acquisition instead of sending public channel feedback [10]. This will make Eve ignorant to the indices of the subcarriers used for data transmission.

Motivated by these facts, secrecy performance of the proposed design can be enhanced when TDD is adopted by utilizing the remaining nulled subcarriers, which are not used for data transmission. Particularly, these nulled subcarriers can be injected and filled by well-designed artificially interfering signals (AIS) that can only degrade Eve's reception as visualized in Fig. 3 while not affecting Bob's performance whatsoever.

It is obvious from Fig. 3 that, with respect to Bob, the transmitted data points correspond to high subchannel gains, while the AIS-injected subcarriers correspond to deep-faded subchannels; on the other hand, it is not the same with respect to Eve, whose channel looks random with respect to the optimally selected subcarriers at Alice. We call this secure transmission scheme enhanced OFDM-subcarrier index selection with artificially interfering signals (OFDM-SIS-AIS) injection. In this scheme, the transmitted baseband signal by Alice can be formulated as

$$\mathbf{x} = \mathbf{C}\mathbf{F}^{\mathbf{H}}\mathbf{R}\mathbf{P}\begin{bmatrix}\mathbf{s}_d\\ \mathbf{s}_r\end{bmatrix} \in \mathbb{C}^{[(N+L)\times 1]}, \tag{9}$$

where $\mathbf{s}_d = [s_{(1)}, s_{(2)}, \ldots, s_{(N-N_r)}]^T$ is a vector of $N_d = N - N_r$ frequency data symbols and $\mathbf{s}_r = [s_{(1)}, s_{(2)}, \ldots, s_{(N_r)}]^T$ is a vector of N_r artificially interfering symbols within each OFDM block.

In the proposed design, $\mathbf{s}_r$ is judiciously designed to have a distribution and power level similar to that of the data-modulated symbols. This is made as so in order to preclude and prohibit Eve from distinguishing the samples of AIS vector ($\mathbf{s}_d$) from that of the data vector itself ($\mathbf{s}_r$). For instance, in the case when QPSK modulation is used, $\mathbf{s}_r$ can be designed according to the below formula

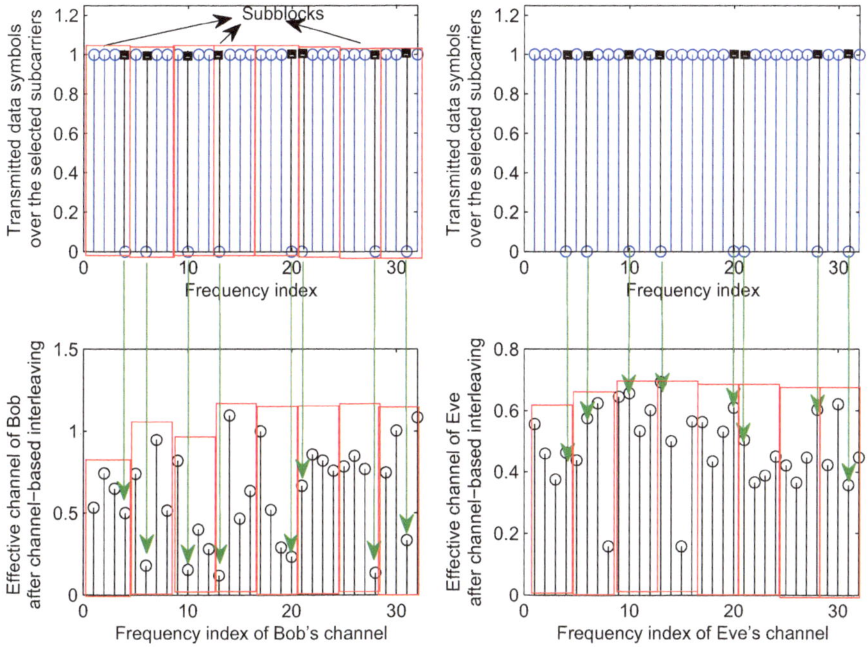

Fig. 3 Subcarrier structure of the designed secure OFDM-SIS-AIS scheme with $\zeta = 3/4$: In each subblock, surrounded by red rectangle, the subcarriers (blue-colored) experiencing good subchannel gains with respect to Bob are used for data transmission, while the rest (black-colored) are injected with AIS. Note that with respect to Eve, the AIS-injected subcarriers do not usually correspond to her weak (bad) subchannels as opposed to Bob's subchannels

$$\mathbf{s}_r = \sqrt{\frac{1}{2}}\left((2\mathbf{u} - 1) + j(2\mathbf{q} - 1)\right) , \tag{10}$$

where the samples of **u** and **q** vectors are chosen to be Bernoulli-distributed random variables with values of ones and zeros. Note that the positions of the AIS are channel-dependent and thus cannot be known by Eve when TDD system is used [14]. For Bob, the positions (indices) of these AIS can be obtained from the previous optimization algorithm as the complement of the indices that maximize the SNR at Bob.

Accordingly, the captured frequency domain signal at Bob's side can be given as

$$\mathbf{y_b} = \mathbf{H_b^f R P} \begin{bmatrix} \mathbf{s}_d \\ \mathbf{s}_r \end{bmatrix} + \hat{\mathbf{z}}_\mathbf{b} \in \mathbb{C}^{[N \times 1]}. \tag{11}$$

After discarding the subcarriers injected with AIS, $\mathbf{y_b}$ will have the size of $(\zeta N) \times 1$. Bob then employs the low-complexity zero-forcing frequency-domain equalization and deinterleaving to detect the transmitted data symbols.

At Eve's side, its captured signal can be given by

$$\mathbf{y_e} = \mathbf{H_e^f R P} \begin{bmatrix} \mathbf{s}_d \\ \mathbf{s}_r \end{bmatrix} + \hat{\mathbf{z}}_\mathbf{e} \in \mathbb{C}^{[N \times 1]}. \tag{12}$$

Note that Eve cannot avoid the effect of the added AIS in the deep fades of the legitimate channel, which is different from her channel, and thus Eve can only make random guesses. Without loss of generality, for Eve to detect the transmitted symbols, she equalizes its received data symbols vector by its corresponding effective channel frequency response. Note that Eve will not have the same performance as that of the legitimate receiver due to the fact that her channel is different from Bob's one. In other words, since the selected subcarriers at Alice are independent of Eve's channel, the M strongest, selected transmit subcarriers for Bob correspond to a random set selection of transmit subcarriers with respect to Eve.

It is worth mentioning that the extra degree of freedom formed by OFDM-SIS-AIS scheme can provide flexibility in the OFDM design in the sense that it can be exploited to not only enhance secrecy with minimal capacity reduction but also to perform other useful functionalities. More precisely, the AIS can intelligently be redesigned to reduce peak-to-average power ratio (PAPR), out-of-band emission (OOBE), and/or adjacent channel interference (ACI) in multiuser scenario as is the case in unique-word OFDM waveform [30]. These kind of designs are beyond the scope of this paper but can be considered as a future work on the proposed technique from a waveform design perspective.

5 Performance Analysis

Assuming Eve is fully aware and knowledgeable of the applied PHY security technique (OFDM-SIS-AIS), then she can perform any of the following reception procedures: (1) Receive the whole OFDM symbol, and try to randomly guess the indices used for carrying useful data from that used for AIS injection. Note that since the structure of the AIS is the same as that of the data, Eve cannot benefit from using intelligent detection technique and thus can never be certain about which subcarrier indices used for what purpose (i.e., data transmission or AIS injection). (2) Receive the OFDM symbol, and try to guess the indices used for data from that used for AIS injection based on Eve's channel quality in a similar manner as Bob does. However, since the channel of Eve is different and independent from that of Bob, Eve will end up making mistakes in the process of selecting and deciding on the subcarrier indices used for data from that used for AIS injection.[1]

[1]Remark: If Eve is assumed to be not aware of the used security technique, Eve may think and assume that this is a normal received OFDM signal since the structure of the interference subcarriers is the same as that of the data, i.e., there is no distinctive features between the

Note that since the process of identifying and detecting the subcarriers injected by AIS is fully random in both of the above-explained reception procedures, the performance results for both cases will be similar (this will be demonstrated in the numerical evaluation results). Accordingly, the investigation and theoretical analysis of one of the cases is sufficient enough to provide and quantify the detailed performance of the proposed OFDM-SIS-AIS technique.

5.1 *Error Probability Associated with Bob and Eve*

Under the proposed OFDM-SIS-AIS design, the BER of Bob is anticipated to be the same as that of OFDM-SIS for any used selection ratio. This is due to the fact that Bob can (by using his channel estimate) simply detect and identify the subcarriers used for data transmission from those used for injecting AIS. Thus, Bob can exclude all these subcarriers filled with AIS in his detection process, resulting in an interference-free detection with a BER performance equivalent to that of OFDM-SIS technique that we have recently analyzed in our work in [27].

On the flip side, the BER performance of Eve will not be the same as that of Bob due to injecting AIS as well as selecting the transmit subcarriers based on Bob's channel. Thus, Eve will not be able to avoid the detrimental effect of AIS as her channel is different from that of Bob. Specifically, the error probability (P_e) of the inability of Eve to correctly detect the subcarriers injected with AIS in each subblock is equal to the product of two error probability events. The first error event denoted by ($P_{e1} = 1 - P_{c1}$) represents the probability that the $K - M$ subcarriers injected with AIS are detected wrongly in each subblock composed of K subcarriers, whereas the second error event denoted by ($P_{e2} = 1 - P_{c2}$) represents the probability that the $K - M$ subcarriers injected with AIS are detected wrongly in the shortened subblock composed of the remaining $K - (K - M)$ subcarriers. Mathematically, this can be given as

$$\begin{aligned} P_e &= P_{e1} \times P_{e2} = (1 - P_{c1}) \times (1 - P_{c2}) \\ &= \left(1 - \frac{(K-M)}{K}\right) \times \left(1 - \frac{(K-M)}{K-(K-M)}\right). \end{aligned} \tag{13}$$

Thus, at high SNR regimes, Eve's average BER can be simply formulated as

$$BER_e = \frac{1}{2} \times P_e = \frac{1}{2} \times \frac{M}{K} \times \frac{2M-K}{M}. \tag{14}$$

---

subcarriers carrying data from that carrying interference at the receiving side with respect to Eve. Hence, Eve will try to decode the whole OFDM symbol as in usual cases without considering the fact that there are subcarriers that are filled with interference. In this case, the analysis will give worse results compared to the abovementioned cases.

For the special case when $\zeta = 3/4$, P_e can be given as

$$\begin{aligned} P_e &= \left(1 - \frac{(4-3)}{4}\right) \times \left(1 - \frac{(4-3)}{4-(4-3)}\right) \\ &= \left(\frac{3}{4}\right) \times \left(\frac{2}{3}\right) = \frac{1}{2}. \end{aligned} \tag{15}$$

In this case when $\zeta = 3/4$, BER_e is given as

$$BER_e = \frac{1}{2} \times P_e = \frac{1}{2} \times \frac{1}{2} = \frac{1}{4}. \tag{16}$$

For the case when $\zeta = 2/4$, $BER_e = \frac{1}{2}$.

5.2 Secrecy Outage Probability

In this subsection, we use the secrecy outage probability as a metric to analytically evaluate the secrecy performance of the proposed OFDM-SIS-AIS scheme. Secrecy outage is chosen as a suitable metric to quantify the performance because of the fact that the CSI of Eve's channel in a practical passive eavesdropping scenario is neither available to Alice nor to Bob. Before calculating secrecy outage, we need first to determine the effective SNR formulas at both Bob and Eve. Besides, we need to determine the effective channel distribution of both the good subchannel used for data transmission and the bad ones used for injecting AIS.

Since Bob knows exactly the locations (indices) of the subcarriers filled with interfering signals, he can just discard these subcarriers carrying non-data interfering signals. Thus, Bob's effective SNR will not be affected by interference, and it can be given per each subblock as

$$\gamma_b = \frac{\zeta |H_{b_i}|^2 P}{\sigma_b^2} \tag{17}$$

$$\gamma_e = \frac{\zeta |H_{e_i}|^2 P}{(1-\zeta)|H_{e_i,int}|^2 P_{int} + \sigma_e^2} \tag{18}$$

$$\gamma_e = \frac{\phi_1 \sigma}{\phi_2 \tau + 1}, \tag{19}$$

where $\phi_1 = \frac{\zeta P}{\sigma_e^2}$, $\phi_2 = \frac{(1-\zeta)P_{int}}{\sigma_e^2}$, $\sigma = |H_{e_i}|^2$, and $\tau = |H_{e_i,int}|^2$. Note that both $|H_{e_i}|^2$ and $|H_{e_i,int}|^2$ have exponential power distribution. Consequently, the effective CDF of γ_e can be found as follows:

$$F(\gamma_e) = \int_{\tau=0}^{+\infty} \int_{\sigma=0}^{\gamma_e(\tau+1)} f(\sigma, \tau) d\sigma \, d\tau \tag{20}$$

$$= \int_{\tau=0}^{+\infty} \int_{\sigma=0}^{\gamma_e(\tau+1)} \phi_1 \phi_2 e^{(-\phi_1 \sigma)} e^{(-\phi_2 \tau)} d\sigma \, d\tau \tag{21}$$

$$= \phi_1 \phi_2 \int_{\tau=0}^{+\infty} e^{(-\phi_2 \tau)} \int_{\sigma=0}^{\gamma_e(\tau+1)} e^{(-\phi_1 \sigma)} d\sigma \, d\tau \tag{22}$$

$$= \phi_2 \int_{\tau=0}^{+\infty} e^{(-\phi_2 \tau)}.(1 - e^{(-\phi_1 (1+\tau).\gamma_e)}) d\tau \tag{23}$$

$$= \phi_2 \left(\frac{1}{\phi_2} - \frac{e^{(-\phi_1 \gamma_e)}}{\phi_2 + \phi_1.\gamma_e} \right). \tag{24}$$

To find the PDF of γ_e, we derive the above CDF with respect to γ_e to obtain

$$f_{\gamma_e}(\gamma_e) = \frac{dF(\gamma_e)}{d\gamma_e} \tag{25}$$

$$= \phi_1 \phi_2 e^{(-\phi_1 \gamma_e)}. \left[\frac{1}{\phi_1 \gamma_e + \phi_2} + \frac{1}{(\phi_1 \gamma_e + \phi_2)^2} \right]. \tag{26}$$

Now, having obtained the PDF and CDF of the effective instantaneous SNR at Eve as well as Bob, one can analytically find and calculate the secrecy outage probability as follows. The secrecy outage probability can be given as [10]

$$P_{\text{sout}} = \Pr\{R_{sec} < R_s\}, \tag{27}$$

where R_{sec} is the instantaneous secrecy rate of the proposed OFDM-SIS-AIS technique and is given by $R_{sec} = [R_b - R_e]^+$, in which $[q]^+$ denotes $\max\{0; x\}$, $R_b = \log_2(1 + \gamma_b)$ is the instantaneous rate of the Bob's channel, and

$R_e = \log_2(1+\gamma_e)$ is the instantaneous rate of the Eve's channel, whereas $R_s > 0$ is a predefined targeted secrecy rate. The secrecy outage probability can be further defined as [9]

$$\begin{aligned} P_{\text{sout}} &= \Pr[R_{sec} < R_s \mid \gamma_b > \gamma_e] \Pr[\gamma_b > \gamma_e] \\ &+ \Pr[R_{sec} < R_s \mid \gamma_b \le \gamma_e] \Pr[\gamma_b \le \gamma_e]. \end{aligned} \tag{28}$$

Since $\Pr[R_{sec} < R_s \mid \gamma_b \le \gamma_e]$ always equals to unity, the above formula can be reduced to

$$\begin{aligned} P_{\text{sout}} &= \Pr[R_{sec} < R_s \mid \gamma_b > \gamma_e] \Pr[\gamma_b > \gamma_e] \\ &+ \Pr[\gamma_b \le \gamma_e]. \end{aligned} \tag{29}$$

Using probability concepts, we can rewrite the previous formula as

$$P_{\text{sout}} = \int_0^{\infty} F_{\gamma_b}\left(2^{R_s}(1+x) - 1\right) f_{\gamma_e}(x)dx, \tag{30}$$

where $x = \gamma_e$, $f_{\gamma_e}(x) = P_{\gamma_e}(\gamma_e)$ and $F_{\gamma_b}(.)$ is the CDF of Bob. The CDF of γ_b heavily depends on the selection ratio (ζ) as explained in [27]. For the case when $\zeta = 2/4$, Bob's CDF is given as [27]

$$F_{\gamma_b}(x) = G\left(\frac{\sqrt{\pi}\,\text{erf}\left(\sqrt{\rho}\sqrt{x}\right)}{2\rho^{\frac{3}{2}}} - \frac{\sqrt{x}\text{e}^{-\rho x}}{\rho}\right), \tag{31}$$

where erf(.) is the error function [31]. By substituting the CDF and PDF of the effective instantaneous SNR of Bob and Eve, respectively, into (30), one can get the final expression of P_{sout}.

6 Simulation Results

In this section, we provide simulation results to demonstrate and validate the effectiveness of the proposed security schemes and to also examine the impacts of the selection ratio and the average SNRs on the security performance.

We consider a practical SISO-OFDM system with $N = 64$ subcarriers adopting quadrature phase-shift keying (QPSK) modulation and a guard period of length L. The number of subblocks in each OFDM block is considered to be $N/K = 16$, where each subblock contains $K = 4$ subcarriers. Two different values for the selection ratio ζ are considered, i.e., $\zeta = 3/4$ and $\zeta = 2/4$. The channel is modeled as an independent and identically distributed (i.i.d.) block fading, where channel coefficients are drawn from a Rayleigh fading distribution and the channel is deemed to be slowly varying. The Rayleigh multipath fading channels of both Bob and Eve are assumed to have the same length, $L = 9$ samples, with a sparse, normalized power delay profile given by $\mathbf{p} = [0.8407, 0, 0, 0.1332, 0, 0.0168, 0.0067, 0, 0.0027]$ mW.

Additionally, we consider that Eve is aware of the transmission technique used by Alice and also knows her CSI but does not know the channel of the legitimate link. In the performance evaluation, we use BER-based secrecy gap metric [14] to both evaluate the secrecy gap between Bob and Eve and to quantify the amount of information leakage to Eve.

Figure 4 exhibits the BER performance of the benchmark scheme (i.e., OFDM-SIS [27]) for both Bob and Eve. It is assumed that Eve is aware of the used scheme as well as the indices of the nulled subcarriers (which are not used for data transmission in OFDM-SIS) as these subcarriers have very low power (ideally zero), making it easy for her to identify and then exclude them from the detection process. It is shown that Eve's performance is the same as that of standard OFDM. This happens due to the use of channel-dependent optimal subcarrier indices selection with respect to Alice-to-Bob channel that is different from Alice-to-Eve channel, for which the selection process looks random (not optimal). Thus, the system response will not be favorable to Eve, and no performance gain is delivered to her side. In contrast, Bob's BER is enhanced as the selection ratio decreases due to sending the data over

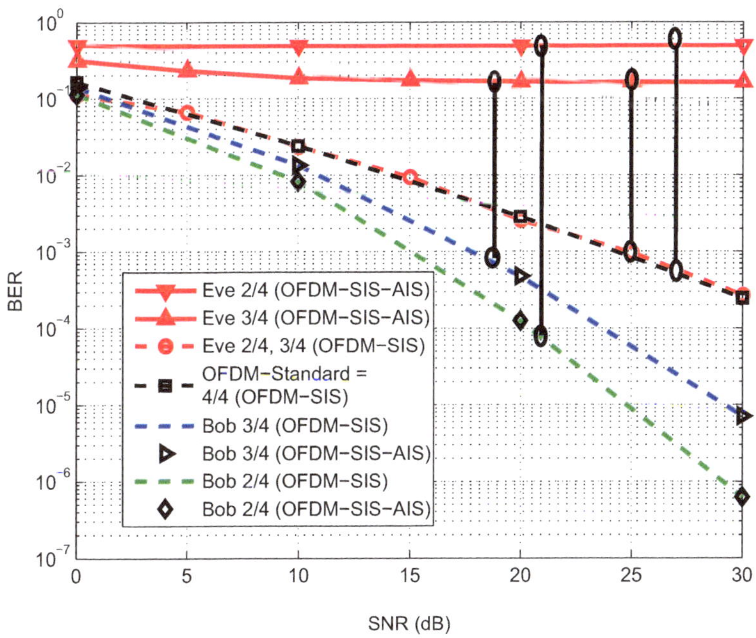

Fig. 4 BER performance comparison between Bob and Eve when OFDM-SIS-AIS is used compared to OFDM-SIS. QPSK modulation alongside $\zeta = 3/4$ and $\zeta = 2/4$ are adopted. TDD mode is considered, where Eve does not know Bob's CSI [14]

the subcarriers that are experiencing good subchannels with respect to only Bob. As noticed, the secrecy gap between Bob's and Eve's channel is promising and can be utilized to provide QoS-based secrecy [20].

However, as it can be observed, secrecy cannot be maintained over all SNR values. For instance, for $\zeta = 3/4$ and for the case when Bob's SNR is equal to or less than 18 dB, while Eve's SNR is equal to or greater than 25 dB, Eve will be able to reliably decode a service of BER requirement equals to 10^{-3}, while Bob cannot. Thus, there is a security breach as the secrecy gap in this case will be negative. This problem can be addressed by the proposed OFDM-SIS-AIS scheme, where positive secrecy gap can be ensured at any SNR Bob and Eve may experience.

Figure 4 presents the BER performance of both Bob and Eve using the proposed OFDM-SIS-AIS scheme with $\zeta = 3/4$ and $\zeta = 2/4$, compared to OFDM-SIS with the same ζ values and standard OFDM (equivalent to OFDM-SIS with $\zeta = 4/4$). It is depicted that the BER performance of Bob using OFDM-SIS-AIS remains the same as that of OFDM-SIS since Bob can just discard the subcarriers not used for data transmission as he knows his CSI, which is assumed in this technique to be estimated using channel sounding techniques in a TDD system. With respect to Eve, a considerable BER degradation is observed over all SNR ranges due to the artificially injected interfering signals, which cannot be avoided by Eve due to having a channel different than that of Bob. It is observed that the secrecy gap as well as the BER performance of Bob increases as the selection ratio decreases.

Figure 5 depicts the secrecy outage probability performance versus Bob's average SNR. The performance achieved by the proposed OFDM-SIS-AIS scheme is also compared with conventional OFDM-SIS and standard OFDM schemes when the predefined secrecy rate threshold is set to unity (i.e., $R_s = 1$) and Eve's average SNR ($\bar{\gamma}_e$) is equal to 10 dB, whereas the selection ratio ζ equals to 2/4 and 3/4. From Fig. 5, we first observe the OFDM-SIS-AIS scheme has superior performance (lower secrecy outage) compared to conventional OFDM-SIS scheme for $\zeta = 3/4$ as well as $\zeta = 2/4$. In addition, we notice that the decrease of ζ yields a better (lower) secrecy outage performance as the effective SNR at Bob increases, whereas the injection of AIS worsens Eve's SNR, resulting in a larger SNR difference between Bob and Eve.

In Fig. 6, we fix the selection ratio ζ at 3/4 and change Eve's average SNR $\bar{\gamma}_e$ from 10 dB (as it was in the previous setup, whose performance is depicted in Fig. 5) to 0 and 20 dB. This is set as so in order to investigate the effect of changing $\bar{\gamma}_e$ on the secrecy outage performance of OFDM-SIS-AIS and how it compares to conventional OFDM-SIS scheme. We can see that the secrecy outage performance gets lower when $\bar{\gamma}_e$ decreases as expected. A more interesting observation is that the performance difference between the proposed OFDM-SIS-AIS scheme and the conventional OFDM-SIS gets larger as $\bar{\gamma}_e$ increases, and vice versa (i.e., it gets lower as $\bar{\gamma}_e$ decreases). The reason for this behavior is that the injection of AIS becomes less effective in degrading Eve's performance when her SNR is already very bad (like 0 dB). Thus, the performance of OFDM-SIS-AIS boils down (i.e., become comparable) to that obtained by OFDM-SIS when Eve's SNR is extremely very low as demonstrated in Fig. 6.

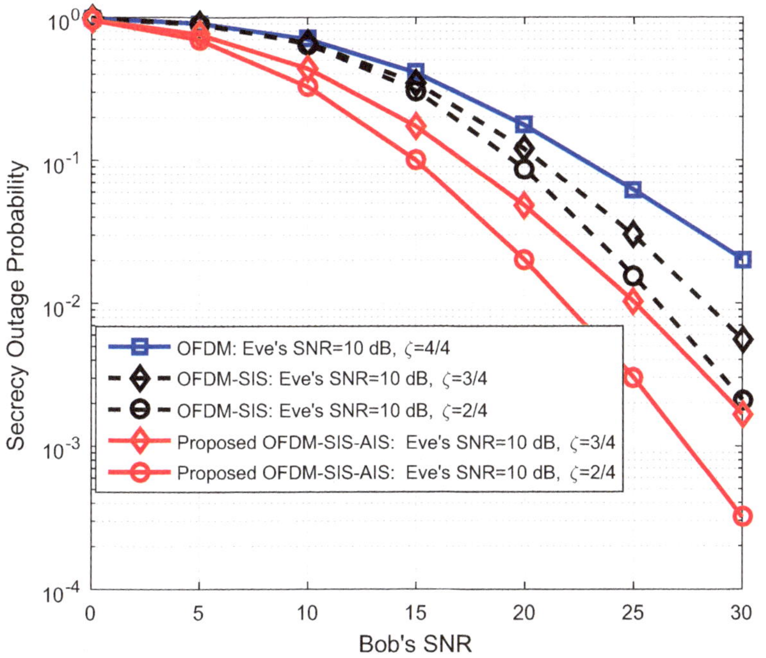

Fig. 5 Secrecy outage probability of the proposed OFDM-SIS-AIS for $\zeta = 1$, 3/4, 2/4, $\bar{\gamma}_e = 10$ dB, and $R_s = 1$ compared to conventional OFDM-SIS

Particularity, Fig. 6 shows that the secrecy outage probability performance of OFDM-SIS-AIS scheme is much better than that obtained by OFDM-SIS scheme when Eve's SNR is high (e.g., 20 dB) where Eve is very close to the transmitter, Alice. This occurs as in this case Eve will be highly affected negatively by the injected artificially interfering signals whose power levels are stronger at shorter distances from the transmitter. On the other hand, the performance difference between OFDM-SIS-AIS and OFDM-SIS schemes becomes very insignificant when Eve's SNR is low (e.g., 0 dB) as in this case the effect of the injected artificially interfering signals whose power levels are weaker at longer distances from the transmitter becomes insignificant on degrading the Eve's performance.

It is noteworthy to mention here that there is a clear trade-off between secrecy and reliability from one side and throughput from another side (i.e., secrecy increases as ζ decreases). However, the new degree of freedom resulted from the controllable selection process can bring more advantages in terms of providing more flexibility to the OFDM design. More precisely, the subcarriers which are used to send AIS because of their low channel gains (which already limit the performance of both BER and throughput) can be deliberately filled with specially optimized signals that can fulfill other important functionalities (besides secrecy) such as reducing PAPR and OOBE.

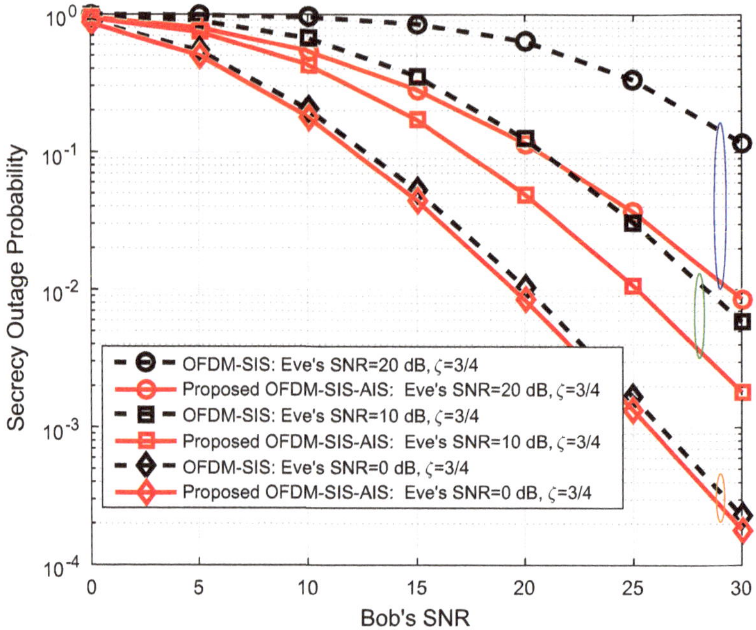

Fig. 6 Secrecy outage probability of the proposed OFDM-SIS-AIS for $\bar{\gamma}_e = 0, 10, 20$ dB, $\zeta = 3/4$ and $R_s = 1$ compared to conventional OFDM-SIS

Based on the obtained results, we have demonstrated that the performance of the proposed OFDM-SIS-AIS outperforms that of OFDM-SIS in terms of secrecy while maintaining the same reliability performance as that of OFDM-SIS.

This enhancement is achieved without sharing secret keys nor knowing Eve's channel nor even causing any major changes in the receiver design. Given the simplicity of the proposed design, its hardware testbed implementation is very handy to structure, making it very attractive for flexible 5G and beyond systems and low-complexity Internet of Things (IoT) devices.

7 Conclusion

An enhanced physical layer security technique has been proposed for safeguarding OFDM-based transmission against eavesdropping. In this technique, named as OFDM-SIS-AIS, the secrecy performance of the recently proposed OFDM-SIS technique is boosted by injecting artificiality interfering signals (AIS) in the subcarriers that are not used for data transmission, resulting in a significant improvement in the secrecy gap when TDD system is used. The presented analytical and simulation results have demonstrated the capability of the proposed schemes in achieving practical secrecy without increasing the complexity of the OFDM structure or knowing Eve's channel, making it very suitable for low-complexity 5G-IoT and Tactile Internet applications. Future work can consider designing and investigating the secrecy performance of different variations of the proposed OFDM-SIS-AIS scheme assuming different subblock sizes and activation ratios.

References

1. Al-Turjman, F., Ever, E., & Zahmatkesh, H. (2018). Small cells in the forthcoming 5G/IoT: Traffic modelling and deployment overview. *IEEE Communications Surveys Tutorials*, 1–1. https://ieeexplore.ieee.org/document/8430735
2. Al-Turjman, F., & Alturjman, S. (2018). Context-sensitive access in industrial internet of things (IIoT) healthcare applications. *IEEE Transactions on Industrial Informatics, 14*(6), 2736–2744.
3. Yang, N., Wang, L., Geraci, G., Elkashlan, M., Yuan, J., & Renzo, M. D. (2015). Safeguarding 5G wireless communication networks using physical security. *IEEE Communications Magazine, 53*(4), 20–27.
4. Mukherjee, A. (2015). Physical-layer security in the internet of things: Sensing and communication confidentiality under resource constraints. *Proceedings of the IEEE, 103*(10), 1747–1761.
5. Al-Turjman, F. (2019). 5G-enabled devices and smart-spaces in social-IoT: An overview. *Future Generation Computer Systems, 92*, 732–744. https://www.sciencedirect.com/science/article/pii/S0167739X17311962
6. Deebak, B. D., Ever, E., & Al-Turjman, F. (2018). Analyzing enhanced real-time uplink scheduling algorithm in 3GPP LTE-advanced networks using multimedia systems. *Transactions on Emerging Telecommunications Technologies, 29*(10), e3443, ETT-18-0041.R2. https://onlinelibrary.wiley.com/doi/full/10.1002/ett.3443
7. Alabady, S. A., Al-Turjman, F., & Din, S. (2018). A novel security model for cooperative virtual networks in the IoT era. *International Journal of Parallel Programming*. https://link.springer.com/article/10.1007/s10766-018-0580-z
8. Al-Turjman, F., & Alturjman, S. (2018). Confidential smart-sensing framework in the IoT era. *The Journal of Supercomputing, 74*, 5187–5198.
9. Mukherjee, A., Fakoorian, S., Huang, J., & Swindlehurst, A. L. (2014). Principles of physical layer security in multiuser wireless networks: A survey. *IEEE Communications Surveys and Tutorials, 16*(3), 1550–1573.
10. Guvenkaya, E., Hamamreh, J. M., & Arslan, H. (2017). On physical-layer concepts and metrics in secure signal transmission. *Physical Communication, 25*, 14–25.
11. Hamamreh, J. M., & Arslan, H. (2018). Joint PHY/MAC layer security design using ARQ with MRC and null-space independent PAPR-aware artificial noise in SISO systems. *IEEE Transactions on Wireless Communications, 17*(9), 6190–6204.
12. Hamamreh, J. M., Guvenkaya, E., Baykas, T., & Arslan, H. (2016). A practical physical-layer security method for precoded OSTBC-based systems. In *Proceedings of 2016 IEEE wireless communications and networking conference (WCNC)*, Apr 2016 (pp. 1–6).
13. Li, M., Kundu, S., Pados, D. A., & Batalama, S. N. (2013). Waveform design for secure SISO transmissions and multicasting. *IEEE Journal on Selected Areas in Communications, 31*(9), 1864–1874.
14. Hamamreh, J. M., & Arslan, H. (2017). Secure orthogonal transform division multiplexing (OTDM) waveform for 5G and beyond. *IEEE Communications Letters, 21*(5), 1191–1194. https://ieeexplore.ieee.org/document/7814269

15. Hamamreh, J. M., & Arslan, H. (2017). Time-frequency characteristics and PAPR reduction of OTDM waveform for 5G and beyond. In *2017 10th international conference on electrical and electronics engineering (ELECO)*, Nov 2017 (pp. 681–685).
16. El Hajj Shehadeh, Y., Alfandi, O., & Hogrefe, D. (2012). Towards robust key extraction from multipath wireless channels. *Journal of Communications and Networks, 14*(4), 385–395.
17. Li, H., Wang, X., & Chouinard, J.-Y. (2015). Eavesdropping-resilient OFDM system using sorted subcarrier interleaving. *IEEE Transactions on Wireless Communications, 14*(2), 1155–1165.
18. Al-Turjman, F. (2018). QoS-aware data delivery framework for safety-inspired multimedia in integrated vehicular-IoT. *Computer Communications, 121*, 33–43.
19. Ng, D. W. K., Lo, E. S., & Schober, R. (2012). Energy-efficient resource allocation for secure OFDMA systems. *IEEE Transactions on Vehicular Technology, 61*(6), 2572–2585.
20. Hamamreh, J. M., Yusuf, M., Baykas, T., & Arslan, H. (2016). Cross MAC/PHY layer security design using ARQ with MRC and adaptive modulation. In *Proceedings of 2016 IEEE Wireless Communications and Networking Conference (WCNC)*, Apr 2016 (pp. 1–7).
21. Hamamreh, J. M., Furqan, H. M., & Arslan, H. (2017). Secure pre-coding and post-coding for OFDM systems along with hardware implementation. In *Proceedings of 2017 13th International Wireless Communications and Mobile Computing Conference (IWCMC)*, June 2017 (pp. 1338–1343).
22. Guvenkaya, E., & Arslan, H. (2014). Secure communication in frequency selective channels with fade-avoiding subchannel usage. In *Proceedings of 2014 IEEE international conference on communications work. ICC'14*, June 2014 (pp. 813–818).
23. Furqan, H. M., Hamamreh, J. M., & Arslan, H. (2017). Enhancing physical layer security of OFDM-based systems using channel shortening. In *Proceedings of 2017 IEEE international symposium on personal, indoor and mobile radio communications (PIMRC)*, Oct 2017 (pp. 8–13).
24. Qin, H., Sun, Y., Chang, T.-H., Chen, X., Chi, C.-Y., Zhao, M., & Wang, J. (2013). Power allocation and time-domain artificial noise design for wiretap OFDM with discrete inputs. *IEEE Transactions on Wireless Communications, 12*(6), 2717–2729.
25. Yusuf, M., & Arslan, H. (2016). Controlled inter-carrier interference for physical layer security in OFDM systems. In *Proceedings of IEEE vehicular technology conference (VTC-Fall)*, Sept 2016 (pp. 1–5).
26. Ankaral, Z. E., Karabacak, M., & Arslan, H. (2014). Cyclic feature concealing CP selection for physical layer security. In *2014 IEEE military communications conference*, Oct 2014 (pp. 485–489).
27. Hamamreh, J. M., Basar, E., & Arslan, H. (2017). OFDM-subcarrier index selection for enhancing security and reliability of 5G URLLC services. *IEEE Access, 5*, 25863–25875. https://ieeexplore.ieee.org/document/8093591
28. Lei, S.-W., & Lau, V. K. N. (2002). Performance analysis of adaptive interleaving for OFDM systems. *IEEE Transactions on Vehicular Technology, 51*(3), 435–444.
29. Basar, E., Wen, M., Mesleh, R., Renzo, M. D., Xiao, Y., & Haas, H. (2017). Index modulation techniques for next-generation wireless networks. *IEEE Access, 5*(1), 16693–16746.
30. Huemer, M., Hofbauer, C., & Huber, J. B. (2010). The potential of unique words in OFDM. In *Proceedings of 15th international OFDM-workshop 2010 (InOWo'10)*, Sept 2010 (pp. 140–144).
31. Gradshteyn, I. S., & Ryzhik, I. M. (2007). *Table of integrals, series and products P.891 (8.258)* (7th ed.). Cambridge, MA: Academic Press. https://www.sciencedirect.com/book/9780123849335/table-of-integrals-series-and-products

Emotional ANN (EANN): A New Generation of Neural Networks for Hydrological Modeling in IoT

Vahid Nourani, Amir Molajou, Hessam Najafi, and Ali Danandeh Mehr

1 Introduction

Referring to the water resources engineering literature, it is observed that artificial intelligence (AI) methods are applied as versatile decision support tools to solve wide range of associated problems. They were typically implemented to attain the cause-effect relationships between nonlinear hydro-environmental processes (i.e., system identification) and time series modeling hydro-meteorological variables that often cannot be modeled by classic statistical models such as autoregressive integrated moving average with exogenous input (ARIMAX). A few studies are also available using AI methods as system identifiers of hydrological processes (e.g., [3, 4, 10, 11, 36, 38, 57]).

Most of the hydrological processes are known as highly nonlinear process that cannot be expressed in simple or complex mathematical forms [48]. To address such difficulty, AI-based modeling can be more effective than the probabilistic or distributed (physically based) models mainly because of (i) complex underling systems of the hydrological processes, (ii) unknown factors/parameters involved in the processes, (iii) and spatiotemporal variation of the processes and their forcing

V. Nourani · H. Najafi
Department of Water Resources Engineering, Faculty of Civil Engineering, University of Tabriz, Tabriz, Iran

A. Molajou
Department of Water Resources Engineering, Faculty of Civil Engineering, Iran University of Science & Technology, Tehran, Iran

A. Danandeh Mehr (✉)
Department of Civil Engineering, Faculty of Engineering, Antalya Bilim University, Antalya, Turkey
e-mail: ali.danandeh@antalya.edu.tr

F. Al-Turjman (ed.), *Artificial Intelligence in IoT*,
Transactions on Computational Science and Computational Intelligence,
https://doi.org/10.1007/978-3-030-04110-6_3

factors. Some of the AI methods applied in hydrological studies include (but not limited to) artificial neural networks (ANNs, e.g., [1, 34, 35]), fuzzy logic (FL, e.g., [36, 45]), support vector regression (SVR, e.g., [8, 12]), and genetic programming (GP, e.g., [11, 27]).

ANNs, known as the universal approximator, have been widely used for modeling nonlinear hydrological processes over the past decades [5]. In general, the advantages of ANN models in comparison with other statistical and conceptual methods can be categorized as [39]:

- Regarding to the black box feature of ANN models, the use of these models does not require prior knowledge of the process.
- Due to the application of a nonlinear filter, known as the activation function on neurons, ANN models can handle the nonlinear properties of the process.
- ANN models have the ability to apply multivariate inputs with different characteristics.

The ability of ANN for linking input and output variables in complex hydrological systems without the need of prior knowledge about the nature of the process has led to a huge leap in the use of ANN models in hydrological simulations [2, 5, 13, 14, 22, 23, 32, 42].

In spite of popularity and capability of nonlinear modeling, ANNs suffer from some deficiencies when the interested hydrological process includes inadequate observed samples or the associated time series comprises high rate of non-stationary and seasonal variations [39]. To achieve reliable models and increase the accuracy of results, a number of data preprocessing approaches such as wavelet transform, season algorithm, singular spectrum analysis, and others have been developed and used in the hydrological modeling issues [8, 12, 39, 49]. The effectiveness of wavelet-based de-noising and multi-resolution analysis in optimizing AI models has been recently introduced and widely employed by the hydrologists to simulate different components of the hydrologic cycle such as rainfall-runoff, river flow, groundwater, precipitation, and sedimentation [41]. For example, Kisi and Cimen [26] examined the efficiency of wavelet-SVR to one-day-ahead rainfall predicting in Turkey and demonstrated that the hybrid model can increase forecasting precision and performs superior than the stand-alone SVR and ANN models. Wavelet-based data preprocessing approach has also been employed to extract the seasonal features of the hydrological processes by decomposing the main time series into multi-scale sub-series, each representing a specific seasonal scale [3, 28]. Such studies showed that the data preprocessing by wavelet transform may improve the modeling efficiency over different time scales (both short and long terms). Corresponding improvement was found to be more sensible in large time scales such as seasonal or monthly, because in most of hydrological process, the seasonal (periodic) patterns in the large-scale time series are more dominant than that of the small-scale time series. In other words, the autoregressive property is more remarkable in small-scale hydrological time series (e.g., daily), whereas the seasonal specification is more highlighted in large-scale time series (e.g., monthly) (see [26, 49, 53]). However, it should be noted that such wavelet-based data processing scheme should be conducted within an external unit apart from the ANN's framework.

2 Emotion in ANNs

Fellous is probably the first researcher who stressed on the need for emotions in AI systems describing that the emotions must be dynamically interacted with together [18]. Emotion tends to be used in medical terminology for what a person is feeling at a given moment. Examples for emotion include joy, sadness, anger, fear, disgust, surprise, pride, shame, regret, and elation. The first six emotions are typically considered as the basic emotions, and the last four are treated as elaborations or specializations of them [6]. More recent descriptions either emphasized the external stimuli that trigger emotion or the internal responses involved in the emotional state, when in fact emotion includes both of those things and much more [25]. Perlovsky [43] defines emotion as the exaggeratedly expressive communications related to feelings. Love, hate, courage, fear, joy, sadness, pleasure, and disgust can all be described in both mental and physical terms. Emotion is the realm where thought and physiology are inextricably entwined, and where the self is inseparable from individual perceptions of value and judgment toward others and ourselves [6]. Emotions are sometimes considered as the antithesis of reason. A distinctive and challenging fact about human beings is a potential for both opposition and entanglement between will, emotion, and reason [25]. According to Khashman [25] researchers and scientists studied the role of emotions in artificial intelligence (AI) from a variety of viewpoints: to develop agents and robots that interact more gracefully with humans, to develop systems that use the analog of emotions to aid their own reasoning, or to create agents or robots that more closely model human emotional interactions and learning [29]. Although computers do not have physiologies like humans, information signals and regulatory signals travel within them. According to Picard [44], "There will be functions in an intelligent complex adaptive system, that have to respond to unpredictable, complex information that play the role that emotions play in people. Therefore, for computers to respond to complex affective signals in a real-time way, they will need something like the systems we have, which we call emotions." Such computers will have the same emotional functionality, but not the same emotional mechanisms as human emotions [25, 44].

Recent studies have shown that scientists attempt to integrate the artificial emotion into the ANN in order to solve complex engineering problems via emotional ANN (EANN) models. From the biological standpoint, the mood and emotion of animal due to the activity of hormone glands can affect neurophysiological response of the animal, sometimes by providing different actions for a similar task at different moods [37]. Similarly for an EANN, there will be a feedback loop between the hormonal and neural systems; each is influenced by the other which in turn, the learning ability of the network is relatively enhanced. Over the past decades, a few kinds of EANNs have been developed and suggested. They have their own merits and features. For instance, Moren [33] proposed brain emotional learning (BEL)-based ANN inspired by some biological evidences that an emotional stimulus (such as fear) can be processed more quickly than a regular stimulus through available shorter routes in the emotional brain to act fast when the logical mind does not have enough time for processing an external situation (such as danger).

BEL networks have been efficiently used by Rahman et al. [46] to control interior permanent magnet synchronous motor drive. Khashman [25] considered emotional anxiety and confidence factors to modify back propagation (BP) learning algorithm of the multilayer perceptron (MLP) networks. The author developed emotional back propagation (EmBP) neural network in which anxiety factor was initialized according to the pattern of input samples, and then it was modified through the process of iteration. In a contrary manner, confidence factor was related to the anxiety factor as well as the network output at the first iteration. In the beginning of the network training, the anxiety and confidence levels were found high and low, respectively, but they received optimal values after a few iterations. Within the training procedure of EmBP, assigning a high value to the anxiety factor forces the network to have less attention to the derivative of the errors (error gradient) in the network's output. However the rise of confidence factor (due to stress reduction) dictates the network to pay more heed to the alteration of the weights in the previous training step. In fact, the procedure was similar to the magnification of inertia term to moderate the alteration degree from a pattern to the other as the learning iteration is progressed [37]. In the studies by Lotfi and Akbarzadeh [30, 31], BEL, EmBP, and some other emotional concepts were conjugated in order to develop some EANNs for clustering, pattern recognition, and predication tasks. From the mathematical perspective and apart from the biological concepts, with regard to the conventional ANN, an EANN includes a few extra parameters which are dynamically interacted with inputs, outputs, and statistical weights of the network [37]. Returning to the hydro-environmental studies, Nourani [37] demonstrated the first application of EANN in which the author proposed the revised BP algorithm to train MLP networks by incorporating emotional anxiety concept. The new algorithm was used to solve streamflow forecasting problem when there is lack of long-observed training time series. Details of this pioneer study are described in the following section after a brief overview on the structure of EANN and its difference with classic ANN.

3 Difference Between EANN and Simple ANN

Feed-forward neural networks (FFNN) are of the most popular ANN structures extensively applied to model different components of the hydrologic cycle [4, 13]. A FFNN with three layers of input, output, and hidden, trained by BP algorithm, has shown appropriate efficiency in nonlinear hydrological modeling tasks [5, 20].

Figure 1 shows a schematic of a three-layered FFNN, and the explicit equation to calculate the target of such FFNN can be written as [40]:

$$\widehat{y}_j = f_j\left[\sum_{h=1}^{m} w_{jh} \times f_h\left(\sum_{i=1}^{n} w_{hi}x_i + w_{hb}\right) + w_{jb}\right] \tag{1}$$

where *i*, *h*, *j*, *b*, and *w* indicate, respectively, neurons of the input, hidden, and output layers and bias and applied weight (or bias) by the neuron; f_h and f_j show activation functions of hidden layer and output layer, respectively; x_i, *n*, and *m* represent, respectively, input value, input, and hidden neuron numbers; and *y* and $\widehat{y}_j$denote the observed and calculated target values, respectively. In the calibration phase of the model, the values of hidden and output layers and corresponding weights could be varied and calibrated.

On the other hand, an EANN model is the improved version of a conventional ANN including an emotional system which emits artificial hormones to modulate the operation of each neuron, and in a feedback loop, the hormonal parameters are also adjusted by inputs and output of the neuron. The schematic of an inner neuron from FFNN and EANN has been depicted in Figs. 1 and 2, respectively.

By comparing these two neurons, it is deduced that in contrast to the FFNN in which the information flows only in the forward direction, a neuron of EANN can reversibly get and give information from inputs and outputs and also can provide hormones (e.g., H_a, H_b, and H_c). These hormones as dynamic coefficients are initialized according to the pattern of input (and target) samples and then are modified through the training iterations. Through training phase they can impact on all components of the neuron (i.e., weights, I; net function, II; and activation

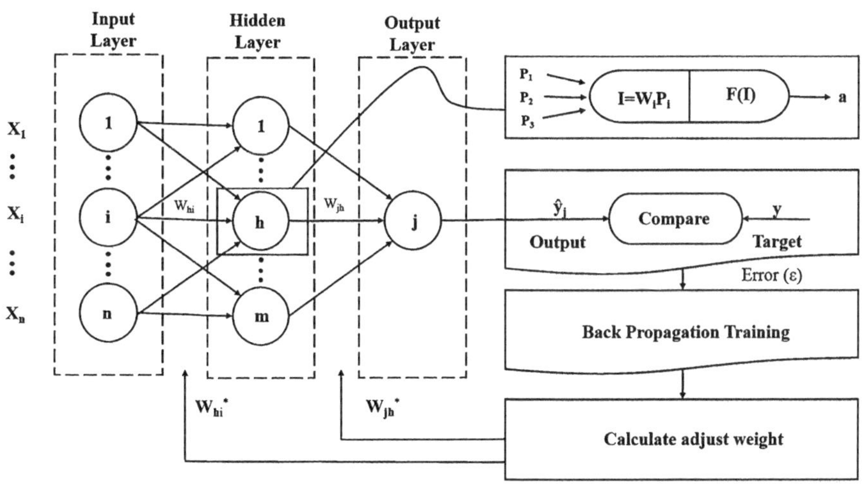

Fig. 1 Schematic of a three layer FFNN model

Fig. 2 A node of EANN and emotional unit [37]

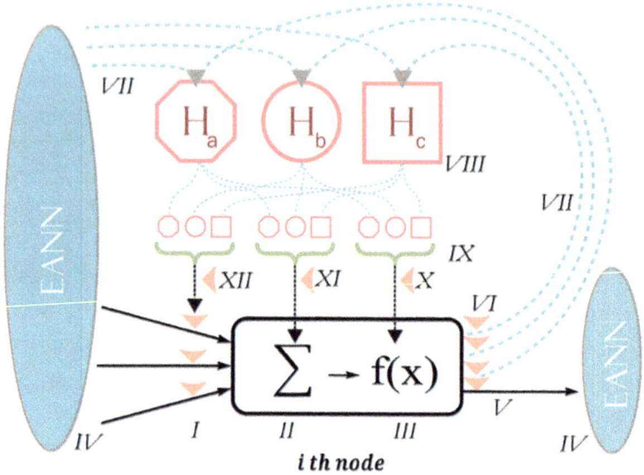

function, III, in Fig. 2). In Fig. 2 the solid and dotted lines, respectively, show neural and hormonal routes of information. The output of ith neuron in an EANN with three hormonal glands of H_a, H_b, and H_c can be computed as [37]:

$$Y_i = \underbrace{\left(\gamma_i + \sum_h \partial_{i,h} H_h\right)}_{1} \times f\left(\sum_j \left[\underbrace{\left(\beta_i + \sum_h \chi_{i,h} H_h\right)}_{2} \times \underbrace{\left(\alpha_{i,j} + \sum_h \Phi_{i,j,k} H_h\right) X_{i,j}}_{3}\right) + \underbrace{\left(\mu_i + \sum_h \psi_{i,h} H_h\right)}_{4}\right]) \tag{2}$$

where the artificial hormones are computed as [37]:

$$H_h = \sum_i H_{i.h} \qquad (h = a, b, c) \tag{3}$$

In Eq. 2, term (1) shows the imposed weight to the activation function (f). It includes the statistic (constant) neural weight of γ_i as well as dynamic hormonal weight of $\sum_h \partial_{i,h} H_h$. Term (2) stands for the imposed weight to the summation (net) function. Term (3) shows imposed weight to the $X_{i,j}$ (an input from jth node of former layer), and term (4) shows the bias of the summation function, including both neural and hormonal weights of μ_i and $\sum_h \psi_{i,h} H_h$, respectively.

The contribution of overall hormonal level of EANN (i.e., H_h) among the hormones should be controlled by $\partial_{i,h}$, $\chi_{i,h}$, $\Phi_{i,j,k}$ and $\psi_{i,h}$ factors which in turn, the i th node output (Y_i) will provide hormonal feedback of $H_{i,h}$ to the network as [37]:

$$H_{i,h} = glandity_{i,h} \times Y_i \tag{4}$$

where the *glandity* factor should be calibrated in the training phase of EANN to provide appropriate level of hormone to the glands. Different schemes may be used to initialize the hormonal values of H_h according to the input samples, e.g., mean of input vector of learning samples. Thereafter considering the network output (Y_i) and Eqs. 2 and 3, the hormonal values are updated through the learning process to get appropriate match between observed and computed time series of the target.

4 Application of EANN for Hydrological Modeling

As previously mentioned, Nourani [37] proposed EANN model to simulate rainfall-runoff process in two watersheds, namely, Lobbs Hole Creek in Australia and Moselle River in France. The study incorporated emotional anxiety into a FFBP structure and investigated its efficiency using different efficiency measures including root-mean-square error (RMSE), determination coefficient (DC), and determination coefficient for peak values (DC_{peak}). To assess the capability of modeling with limited observed time series, the author developed three different scenarios (i.e., strategy1, strategy2, and strategy3) in which the EANN was trained and validated by different size of training and validation samples.

The first case study was demonstrated using observe data from Lobbs Hole Creek, a sub-basin of Murrumbidgee in Australia, and the second was shown using data from Moselle River catchment, a sub-basin of river Rhine in France. The selected watersheds had two distinct climatic conditions showing quite different hydrological behavior and response to the rainfall. The Lobbs Hole Creek is upstream of the main river including mountainous and hilly regions and experiencing irregular precipitation pattern over a year with higher coefficient of variation (because of higher fluctuations of the observed data). However, Moselle watershed experiences a well-dominated seasonal weather with a larger area.

Different data division strategies were assumed to evaluate the overall performance and also efficiency of models to estimate peak discharge values. To this end, the first 75%, 50%, and 40% of data samples were considered as training data sets at the strategies1, strategies2, and strategies13, respectively. By decreasing the number of training samples, the generalization of the resultant network to predict the unseen data sets was foun d as a challenging task. Furthermore, due to importance of multi-step-ahead forecasts of the hydro-environmental processes, already pointed out by some studies (e.g., [7]), 2-, 4-, and 7-day-ahead predictions of the process were performed and compared as well as single-step-ahead forecasting scenario.

The rainfall value at time t (R_t) and current and antecedent runoff values (Q_t, Q_{t-1}, ..., Q_{t-p}) were imposed to the FFBP and EANN to predict runoff value one-d-ahead (Q_{t+1}) as the network output. Since the effect of antecedent rainfall values is implicitly considered by antecedent runoff values, only R_t was entered to the networks as a potential input. Therefore the general mathematical formulation of the networks could be expressed as:

$$Q_{t+k} = f_n\left(Q_t, Q_{t-1}, Q_{t-2}, \ldots, Q_{t-p}, R_t\right) \tag{5}$$

in which f_n denotes the FFBP or EANN and $k = 1$ for single-step-ahead forecasting and $k = 2$, 4, and 7 for multi-step-ahead forecasts.

Selection of appropriate architecture of the network, i.e., appropriate lag No. of p (No. of neurons in input layer will be p + 1), hidden layer No., and optimal iteration epoch No., is a key issue in any training task which can prevent the network from the overtraining problem. Considering tangent sigmoid and pure line as activation functions of hidden and output layers, respectively, the FFBPs were trained using Levenberg-Marquardt scheme of BP algorithm [19], and the best structure and epoch number of each network were determined through trial-error procedure. Also for EANN model, hormones as dynamic coefficients are initialized according to the pattern of input (and target) samples and then are modified through the training iterations. Through training phase, they can impact on all components of the neuron (i.e., weights, net function, and activation function).

Overall comparison of the results denotes superiority of the EANN over FFBP in rainfall-runoff modeling of both watersheds. According to higher coefficient of variation (because of higher fluctuations of the observed data) for the observed rainfall and runoff time series of the Lobbs Hole Creek watershed due to the hydro-geomorphologic condition, this watershed has a wild hydro-climatic regime with regard to the Moselle watershed. Therefore for Lobbs Hole Creek watershed, the performance of both models is lower as compared to the Moselle watershed. However this difference between modeling performances of two watersheds using EANN is relatively lower than the FFBP modeling. The results are listed below:

1. Computed values of DC_{peak} (see Table 1) indicate the ability of EANN to catch the peak values of hydrographs better than FFBP in both watersheds. Since autoregressive models, according to the Markovian property of a process, consider states of the process at some previous time steps to predict the state of system at the next time step, they usually underestimate the peak values which occurred due to instantaneous imposition of an external force (intensive rainfall) to the system. In this condition, the system is experiencing an emotional situation which is different from normal conditions of the system. Therefore in the training phase, a hormone of the emotional unit of EANN acts as a dynamic weight to recurrently gives the feedback to other components of the network and regulates the model for the emotional situation. In the mathematical point of view, such dynamical weights are activated in extraordinary situations (e.g., intensive rainfall) and affect and magnify the weights of network, all done within the EANN framework without any need to external data processing approach.
2. The obtained results by data division strategies of 2 and 3 presented in Table 1 showed that EANN can be trained efficiently even in the presence of the sparse training data sets. In the Lobbs Hole River, the strategies 2 and 3 reulted in 1.4%

Table 1 Results of FFBP and EANN models for both watersheds via three data division strategies

Watershed	Model	Data division strategy	Input[a]	No. hidden neuron	Epoch	DC		RMSE (normalized)		DC Peak
						Train	Verify	Train	Verify	Verify
Lobbs Hole Creek	FFBP	75–25	Q(t), Q(t−1), R(t)	6	80	0.701	0.648	0.122	0.203	0.46
		50–50		5	40	0.682	0.6	0.141	0.23	0.43
		40–60		4	60	0.741	0.454	0.11	0.314	0.41
	EANN	75–25	Q(t), Q(t−1), R(t)	5	20	0.791	0.74	0.091	0.184	0.67
		50–50		5	70	0.764	0.74	0.103	0.182	0.63
		40–60		5	60	0.778	0.7	0.107	0.212	0.61
Moselle	FFBP	75–25	Q(t), Q(t−1), Q(t−2), R(t)	11	90	0.872	0.85	0.0037	0.0038	0.81
		50–50		7	40	0.851	0.823	0.004	0.0041	0.79
		40–60		6	70	0.89	0.8	0.0032	0.0042	0.79
	EANN	75–25	Q(t), Q(t−1), Q(t−2), R(t)	3	20	0.931	0.92	0.002	0.0024	0.9
		50–50		10	10	0.93	0.91	0.0022	0.0024	0.82
		40–60		5	50	0.952	0.901	0.0018	0.0026	0.84

[a]Output is $Q_{(t+1)}$

and 5.7% reduction in the performance of the EANN. Regardin the results of the same strategies in the Moselle Rivere, 1% and 2.2% performance reduction were obtained. Based on the FFBP modelling results, the performance reduction values were 8.5% and 32% for the Lobbs Hole River and 3.5% and 6.3% for the Moselle River. The FFBP even shows overtraining alert for Lobbs Hole Creek watershed in strategy 3. However during FFBP training, the network does not consider the difference between the system's situations, and therefore when the statistical weights of network start to be trained using the ordinary situation data of system (e.g., base flow data of river, observed at the outlet of watershed), sudden appearance of severe rainfall values in the input layer can alter the trained weights, and then again this confusion continues by returning the system to the ordinary state. This is the main reason that usually FFBPs need long data set to be trained appropriately. The aforementioned obtained results approve the efficiency of EANN with regard to conventional FFBP model when it is trained using relatively fewer data samples.

3. Although both FFBP and EANN models are interpolators and must experience critical and extreme conditions in the training phase, the results indicated that EANN is capable of producing more reliable predictions for the unseen data samples. In other words, ENN can generate better estimations using less observed extreme values.
4. Since the ability of a forecasting model to provide a useful horizon of forecasts is a crucial task in hydrological modeling, several lead times of runoff time series (i.e., 2, 4, and 7 d of lead time) were also considered as the networks' outputs to evaluate and compare the performance of FFBP and EANN models in multi-step-ahead runoff forecasts. The results of multi-step-ahead forecasting obtained via FFBP and EANN models with data division strategies of 1 and 3 have been presented in Table 2 for both watersheds. It should be noticed that the outputs of multi-step-ahead models were Q_{t+2}, Q_{t+4} and Q_{t+7}, and only the results of the best networks have been presented in the table. As the results show, by increasing the forecast horizon, the performance of both FFBP and EANN models is decreased mostly due to the magnification of the forecast noise at each forecasting time step. However again, it is clear that the performance of EANN in multi-step-ahead forecasting is relatively better than FFBP model about 10% in average. According to the presented results in Table 2, although by decreasing the number of training samples the training efficiency is increased, the verification performance for unseen data is remarkably decreased, and the difference between calibration and verification DCs is increased.

At the first glance, it brings to mind that EANN is structurally more complicated than FFBP, but actually the EANN with only a few hormonal parameters could lead to better outcomes without the need for any external data processing operation. Even though, in some cases the best structure of trained EANN contains fewer input neurons than FFBP which makes the EANN more comparable with the FFBP model from structural simplicity point of view (see Table 1).

Table 2 Results of multi-step ahead modeling via data division strategies 1 and 3

				DC strategy 1		DC strategy 3	
Watershed	Model	Output	Input	Train	Verify	Train	Verify
Lobbs Hole Creek	FFBP	Q(t + 2)	Q(t), Q(t−1), R(t)	0.7	0.617	0.72	0.608
		Q(t + 4)		0.678	0.611	0.705	0.603
		Q(t + 7)		0.65	0.554	0.672	0.525
	EANN	Q(t + 2)	Q(t), Q(t−1), R(t)	0.775	0.724	0.784	0.708
		Q(t + 4)		0.742	0.712	0.763	0.681
		Q(t + 7)		0.701	0.68	0.717	0.601
Moselle	FFBP	Q(t + 2)	Q(t), Q(t−1), Q(t−2), R(t)	0.828	0.802	0.835	0.778
		Q(t + 4)		0.77	0.751	0.791	0.7
		Q(t + 7)		0.715	0.702	0.723	0.645
	EANN	Q(t + 2)	Q(t), Q(t−1), Q(t−2), R(t)	0.918	0.9	0.921	0.845
		Q(t + 4)		0.87	0.849	0.88	0.822
		Q(t + 7)		0.821	0.788	0.823	0.74

Briefly, through the comparison of the proposed EANN and conventional FFBP models, two main objectives were targeted. Firstly to address the deficiency of network training in the lack of long training time series, three data division strategies with different sizes of training points were considered for the training purpose. The outcomes showed the ability of EANN to cope with the lack of long observed data used for network training. Secondly in the multi-step-ahead forecasting task, the obtained results indicated better performance of EANN than FFBP so that for 2, 4, and 7 d forecasts via EANN model, the reductions of forecasting performance of test data with regard to single-step-ahead forecasting were 2%, 4%, and 8.2% and 2.1%, 7.6%, and 14%, respectively, for the Lobbs Hole and Moselle watersheds. These reductions were 5%, 6.2%, and 15.4% and 5.9%, 11.8%, and 18% for the FFBP model. Overall, the comparison of experimental results shows the merits of EANN in the mentioned tasks of rainfall-runoff modeling with regard to the FFBP model. In contrast to the statistical weights of network, emotional parameters of an EANN dynamically get/give information from/to inputs and outputs of the network at each time step to distinguish the dry (e.g., rainless d) and wet (e.g., stormy d) situations of the system. Both watersheds studied in this research are almost free from remarkable anthropogenic influences. Clearly just like any other data-driven time series forecasting method, the performance of the EANN can be affected in presence of anthropogenic and/or climatic influences and shifts of the observed time series. In the presence of such shifts or strong non-stationary of time series, reliable data preprocessing approaches may be employed prior to performing the forecasts.

Such a reliable implementation of EANN in rainfall-runoff modeling offers its application to model other hydrological processes (e.g., sediment load, groundwater, precipitation, etc.) at different time scales (e.g., daily, monthly, and annual). According to the importance of accurate predictions of hydro-climatologic events (stressed by [55]), the proposed EANN model may be used to create ensemble extreme predictions at multiple lead times. The employed model in this study was a typical form of EANN among broad classes of EANNs trained by BP algorithm; future studies may focus on evaluating other types of EANNs and other training algorithms (e.g., metaheuristic approaches) in hydrological modeling.

More recently Sharghi et al. [51] implemented EANN to model Markovian and seasonal rainfall-runoff process in West Nishnabotna and Trinity Rivers in the USA (sub-basin in California, USA). The authors compared the prediction accuracy of EANN with those of FFNN and wavelet-ANN in terms of different statistical measures and demonstrated that for daily modeling, EANN outperforms the counterparts, especially for the Trinity River. By contrast, the results showed that wavelet-ANN is superior for monthly rainfall-runoff modeling.

5 The Internet of Things (IoT) in Hydro-Environmental Studies

The IoT is a concept in which the virtual world of information technology integrates seamlessly with the real world of things. Many of the initial developments toward the IoTs have focused on the combination of Auto-ID and networked infrastructures in business-to-business logistics and product life cycle applications [54]. Regarding the application of IoT in hydro-environmental studies, our review showed that only a few researches considered the IoT in their studies although web services as an online repository of historical and real-time hydrological data such as runoff, rainfall, streamflow, and groundwater level are available for more than a decade. However, it must be mentioned that geographical information systems (GIS), remote sensing (RS), and data storage systems were applied frequently in the hydro-environmental studies such as flood forecasting, flood hazard mapping, as well as climate change studies (e.g., [9, 50]).

One of the earlier studies in the application of the IoT in the wide range of hydro-environmental studies was carried out by Xiaoying and Huanyan [56]. The authors developed a wetland monitoring system on the basis of real-time, remote, and automatically monitored data in which wireless sensor networks and communication systems were used. The study showed that the new system may provide accurate sampling data that is important for conservation of wetlands. In the preliminary study of possible applications of IoT, Khan et al. [24] reported some applications in hydro-environment such as prediction of natural disasters and water scarcity detection at different places. The combination of sensors and their autonomous coordination and with the relevant modeling tools, one may predict the occurrence of natural disasters and take appropriate actions in advance. In addition, such network may be used to alert the users of a stream or water supply pipelines, for instance, when an upstream event such as the accidental release of sewage into the stream might have dangerous issues for downstream users. In a similar study, Dlodlo and Kalezhi [15] studied the potential applications of the IoT in environmental management in South Africa. The authors categorized IoT applications into four broad classes of environmental quality and protection management, oceans and coasts management, climate change adaptation, biodiversity, and conservation and environmental awareness. The results indicated that integrating IoT into environmental management in South Africa has likely more enhanced impact. Environmental IoT together with 1-year meteorological measurements was employed by Du et al. [16] to investigate the characterization of atmospheric visibility and its relationship with the variables comprising precipitation, relative humidity, wind speed, and wind direction at Xiamen, China. The study demonstrated that an optimal regression model can moderately simulate atmosphere visibility which provides new insights to its characteristics and forcing meteorological factors. Fang et al. [17] focused on the integration of RS data, GIS, and global positioning system with IoT and cloud services to develop snowmelt flood early-warning system for a case study catchment in Xinjiang, China. The results revealed that the process of snowmelt flood simulation and early warning are greatly benefited by such an integrated system. Rathore et al. [47] proposed a hybrid IoT-based system for smart city development and urban planning using Big Data analytics that consists of various types of sensors deployment including smart home sensors, vehicular networking, weather and water sensors, smart parking sensors, surveillance objects, etc. The authors reported that weather and water information may increase the efficiency of the smart city by providing the associated data such as temperature, rain, humidity, pressure, wind speed and water levels at rivers, lakes, dams, and other reservoirs. All the information is gathered by placing the sensors in the reservoirs and other open places. Using rain-measuring sensors and snow-melting parameters, they were able to predict floods and water demands to the residents of the city. More recently, Shenan et al. [52] developed software and hardware in IoT environment to manage light, temperature, and soil water content in a greenhouse system. The authors used FL to monitor and manage the entire process in the system and showed that the single -code fuzzy controllers reside in single microcontroller chip may keep the practicality of the system. Most recently, González-Briones et al. [21] developed an innovative multicomponent system that uses information from wireless sensor networks for knowledge discovery (from weather and terrain conditions) and decision-making in both micro- and macroscale irrigation projects. The use of IoT was improved the efficiency of water use and optimized irrigation system in comparison to a traditional automatic systems.

With respect to the aforementioned review, the present study shows the lack of studies toward the integration of rapidly developing IoT technologies with hydrological modeling techniques, particularly artificial intelligence methods. To increase the efficiency of rainfall-runoff models for many applications in practice, one way may be the integration of IoT with the state-of-the-art EANN that has not been explored so far.

References

1. Abarghouei, H. B., & Hosseini, S. Z. (2016). Using exogenous variables to improve precipitation predictions of ANNs in arid and hyper-arid climates. *Arabian Journal of Geosciences, 9*(15), 663.
2. Abrahart, R. J., Anctil, F., Coulibaly, P., Dawson, C. W., Mount, N. J., See, L. M., et al. (2012). Two decades of anarchy? Emerging themes and outstanding challenges for neural network river forecasting. *Progress in Physical Geography, 36*(4), 480–513.
3. Adamowski, J., Fung Chan, H., Prasher, S. O., Ozga-Zielinski, B., & Sliusarieva, A. (2012). Comparison of multiple linear and nonlinear regression, autoregressive integrated moving average, artificial neural network, and wavelet artificial neural network methods for urban water demand forecasting in Montreal, Canada. *Water Resources Research, 48*(1). https://doi.org/10.1029/2010WR009945.
4. Anmala, J., Zhang, B., & Govindaraju, R. S. (2000). Comparison of ANNs and empirical approaches for predicting watershed runoff. *Journal of Water Resources Planning and Management, 126*(3), 156–166.

5. ASCE Task Committee on Application of Artificial Neural Networks in Hydrology. (2000). Artificial neural networks in hydrology. II: Hydrologic applications. *Journal of Hydrologic Engineering, 5*(2), 124–137.
6. Bonala, S. (2009). *A study on neural network based system identification with application to heating, ventilating and air conditioning (hvac) system* (MSc dissertation). National Institute of Technology, Rourkela.
7. Chang, F. J., & Tsai, M. J. (2016). A nonlinear spatio-temporal lumping of radar rainfall for modeling multi-step-ahead inflow forecasts by data-driven techniques. *Journal of Hydrology, 535*, 256–269.
8. Chau, K. W., & Wu, C. L. (2010). A hybrid model coupled with singular spectrum analysis for daily rainfall prediction. *Journal of Hydroinformatics, 12*(4), 458–473.
9. Danandeh Mehr, A., & Kahya, E. (2017). Climate change impacts on catchment-scale extreme rainfall variability: Case study of Rize Province, Turkey. *Journal of Hydrologic Engineering, 22*(3), 05016037. https://doi.org/10.1061/(ASCE)HE.1943-5584.0001477.
10. Danandeh Mehr, A., Kahya, E., & Olyaie, E. (2013). Streamflow prediction using linear genetic programming in comparison with a neuro-wavelet technique. *Journal of Hydrology, 505*, 240–249.
11. Danandeh Mehr, A., Nourani, V., Hrnjica, B., & Molajou, A. (2017). A binary genetic programing model for teleconnection identification between global sea surface temperature and local maximum monthly rainfall events. *Journal of Hydrology, 555*, 397–406. https://doi.org/10.1016/j.jhydrol.2017.10.039.
12. Danandeh, Mehr, A., Nourani, V., Khosrowshahi, V. K., & Ghorbani, M. A. (2018). A hybrid support vector regression–firefly model for monthly rainfall forecasting. *International journal of Environmental Science and Technology*, 1–12.
13. Danandeh Mehr, A., Kahya, E., Şahin, A., & Nazemosadat, M. J. (2015). Successive-station monthly streamflow prediction using different artificial neural network algorithms. *International journal of Environmental Science and Technology, 12*(7), 2191–2200.
14. Dawson, C. W., & Wilby, R. L. (2001). Hydrological modelling using artificial neural networks. *Progress in Physical Geography, 25*(1), 80–108.
15. Dlodlo, N., & Kalezhi, J. (2015). The internet of things in agriculture for sustainable rural development. In *Emerging Trends in Networks and Computer Communications (ETNCC), 2015 international conference on* (pp. 13–18). IEEE.
16. Du, K., Mu, C., Deng, J., & Yuan, F. (2013). Study on atmospheric visibility variations and the impacts of meteorological parameters using high temporal resolution data: An application of environmental internet of things in China. *International Journal of Sustainable Development & World Ecology, 20*(3), 238–247.
17. Fang, S., Xu, L., Zhu, Y., Liu, Y., Liu, Z., Pei, H., et al. (2015). An integrated information system for snowmelt flood early-warning based on internet of things. *Information Systems Frontiers, 17*(2), 321–335.
18. Fellous, J. M. (1999). Neuromodulatory basis of emotion. *The Neuroscientist, 5*(5), 283–294.
19. Haykin, S. (1994). *Neural networks: A comprehensive foundation*. Upper Saddle River: Prentice Hall PTR.
20. Hornik, K., Stinchcombe, M., & White, H. (1989). Multilayer feedforward networks are universal approximators. *Neural Networks, 2*(5), 359–366.
21. González-Briones, A., Castellanos-Garzón, J. A., Mezquita Martín, Y., Prieto, J., & Corchado, J. M. (2018). A framework for knowledge discovery from wireless sensor networks in rural environments: A crop irrigation systems case study. *Wireless Communications and Mobile Computing, 2018*, 1.
22. Hsu, K. L., Gupta, H. V., & Sorooshian, S. (1995). Artificial neural network modeling of the rainfall-runoff process. *Water Resources Research, 31*(10), 2517–2530.
23. Jain, A., & Srinivasulu, S. (2006). Integrated approach to model decomposed flow hydrograph using artificial neural network and conceptual techniques. *Journal of Hydrology, 317*(3–4), 291–306.

24. Khan, R., Khan, S. U., Zaheer, R., & Khan, S..(2012). Future internet: the internet of things architecture, possible applications and key challenges. In *Frontiers of Information Technology (FIT), 2012 10th International Conference on* (pp. 257–260). IEEE.
25. Khashman, A. (2008). A modified backpropagation learning algorithm with added emotional coefficients. *IEEE Transactions on Neural Networks, 19*(11), 1896–1909.
26. Kisi, O., & Cimen, M. (2011). A wavelet-support vector machine conjunction model for monthly streamflow forecasting. *Journal of Hydrology, 399*(1–2), 132–140.
27. Kisi, O., & Shiri, J. (2011). Precipitation forecasting using wavelet-genetic programming and wavelet-neuro-fuzzy conjunction models. *Water Resources Management, 25*(13), 3135–3152.
28. Kuo, C. C., Gan, T. Y., & Yu, P. S. (2010). Wavelet analysis on the variability, teleconnectivity, and predictability of the seasonal rainfall of Taiwan. *Monthly Weather Review, 138*(1), 162–175.
29. Lewin, D. I. (2001). Why is that computer laughing? *IEEE Intelligent Systems, 16*(5), 79–81.
30. Lotfi, E., & Akbarzadeh-T, M. R. (2014). Practical emotional neural networks. *Neural Networks, 59*, 61–72.
31. Lotfi, E., & Akbarzadeh-T, M. R. (2016). A winner-take-all approach to emotional neural networks with universal approximation property. *Information Sciences, 346*, 369–388.
32. Maier, H. R., & Dandy, G. C. (2000). Neural networks for the prediction and forecasting of water resources variables: A review of modelling issues and applications. *Environmental Modelling & Software, 15*(1), 101–124.
33. Moren, J. (2002). *Emotion and learning: a computational model of theamygdala*, PhD Thesis, Lund university, Lund, Sweden.
34. Moustris, K. P., Larissi, I. K., Nastos, P. T., & Paliatsos, A. G. (2011). Precipitation forecast using artificial neural networks in specific regions of Greece. *Water Resources Management, 25*(8), 1979–1993.
35. Nasseri, M., Asghari, K., & Abedini, M. J. (2008). Optimized scenario for rainfall forecasting using genetic algorithm coupled with artificial neural network. *Expert Systems with Applications, 35*(3), 1415–1421.
36. Nayak, P. C., Sudheer, K. P., Rangan, D. M., & Ramasastri, K. S. (2004). A neuro-fuzzy computing technique for modeling hydrological time series. *Journal of Hydrology, 291*(1–2), 52–66.
37. Nourani, V. (2017). An emotional ANN (EANN) approach to modeling rainfall-runoff process. *Journal of Hydrology, 544*, 267–277.
38. Nourani, V., & Molajou, A. (2017). Application of a hybrid association rules/decision tree model for drought monitoring. *Global and Planetary Change, 159*, 37–45.
39. Nourani, V., Alami, M. T., & Aminfar, M. H. (2009). A combined neural-wavelet model for prediction of Ligvanchai watershed precipitation. *Engineering Applications of Artificial Intelligence, 22*(3), 466–472.
40. Nourani, V., Komasi, M., & Alami, M. T. (2011). Hybrid wavelet–genetic programming approach to optimize ANN modeling of rainfall–runoff process. *Journal of Hydrologic Engineering, 17*(6), 724–741.
41. Nourani, V., Khanghah, T. R., & Baghanam, A. H. (2015). Application of entropy concept for input selection of wavelet-ANN based rainfall-runoff modeling. *Journal of Environmental Informatics, 26*(1), 52–70.
42. Nourani, V., Sattari, M. T., & Molajou, A. (2017). Threshold-based hybrid data mining method for long-term maximum precipitation forecasting. *Water Resources Management, 31*(9), 2645–2658.
43. Perlovsky, L. I. (2006). Toward physics of the mind: Concepts, emotions, consciousness, and symbols. *Physics of Life Reviews, 3*(1), 23–55.
44. Picard, R. W. (1997). *Affective computing*. Cambridge, MA: MIT Press.
45. Pongracz, R., Bartholy, J., & Bogardi, I. (2001). Fuzzy rule-based prediction of monthly precipitation. *Physics and Chemistry of the Earth, Part B: Hydrology, Oceans and Atmosphere, 26*(9), 663–667.

46. Rahman, M. A., Milasi, R. M., Lucas, C., Araabi, B. N., & Radwan, T. S. (2008). Implementation of emotional controller for interior permanent-magnet synchronous motor drive. *IEEE Transactions on Industry Applications, 44*(5), 1466–1476.
47. Rathore, M. M., Ahmad, A., Paul, A., & Rho, S. (2016). Urban planning and building smart cities based on the internet of things using big data analytics. *Computer Networks, 101*, 63–80.
48. Salas, J. D., Delleur, J. W., Yevjevich, V., & Lane, W. L. (1980). *Applied modeling of hydrological time series*. Littleton, CO: Water Resource.
49. Sang, Y. F. (2013). Improved wavelet modeling framework for hydrologic time series forecasting. *Water Resources Management, 27*(8), 2807–2821.
50. Sanyal, J., & Lu, X. X. (2006). GIS-based flood hazard mapping at different administrative scales: A case study in Gangetic West Bengal, India. *Singapore Journal of Tropical Geography, 27*(2), 207–220.
51. Sharghi, E., Nourani, V., Najafi, H., & Molajou, A. (2018). Emotional ANN (EANN) and wavelet-ANN (WANN) approaches for Markovian and seasonal based modeling of rainfall-runoff process. *Water Resources Management, 32*(10), 3441–3456.
52. Shenan, Z. F., Marhoon, A. F., & Jasim, A. A. (2017). IoT based intelligent greenhouse monitoring and control system. *Basrah Journal for Engineering Sciences, 1*(17), 61–69.
53. Shiri, J., & Kisi, O. (2010). Short-term and long-term streamflow forecasting using a wavelet and neuro-fuzzy conjunction model. *Journal of Hydrology, 394*(3–4), 486–493.
54. Uckelmann, D., Harrison, M., & Michahelles, F. (2011). An architectural approach towards the future internet of things. In *Architecting the internet of things* (pp. 1–24). Berlin, Heidelberg: Springer.
55. World Meteorological Organization (WMO), (2012). Guidelines on Ensemble Prediction Systems and Forecasting Report WMO-No. 1091, Switzerland.
56. Xiaoying, S., & Huanyan, Q. (2011). Design of wetland monitoring system based on the internet of things. *Procedia Environmental Sciences, 10*, 1046–1051.
57. Zhang, Q., Wang, B. D., He, B., Peng, Y., & Ren, M. L. (2011). Singular spectrum analysis and ARIMA hybrid model for annual runoff forecasting. *Water Resources Management, 25*(11), 2683–2703.

Smart Tourism Destination in Smart Cities Paradigm: A Model for Antalya

Gözdegül Başer, Oğuz Doğan, and Fadi Al-Turjman

1 Introduction

Emerging technologies influence tourism destinations and cause new challenges arising from changes in both consumers and the environment. The use of technology in a destination can enrich tourist experiences and enhance destination competitiveness as well as enable new distribution channels and create a new business environment [1, 2]. Technology provides interconnection of services as well as supplying information for planning, organizing, and evaluating data. Therefore, technology has emerged as the driving and fundamental force for tourism destinations. Tourism industry is one of the first service industries to adapt and use information and communication technologies (ICT) for promoting its services [2]. However, tourism is not a clear-cut sector but an all-embracing and pervasive domain of service and industrial activities ([3], 2001, p. 5). Tourism services are mostly interconnected and include a wide range of inputs and outputs in tourism destinations. Technology has started to play an important role to interconnect touristic services. As a result, the digital technology has become an important element for the promotion and distribution of tourism's services [4].

In order to deal with these challenges, first destinations have to recognize the kind of changes that occurred and proactively respond to the changes [1]. Implementing technological tools within tourism destinations has become critical since the connected, better informed, and engaged tourists are dynamically interacting with the destination [5]. UNESCO and WTO have certainly indicated that countries need

G. Başer (✉) · O. Doğan · F. Al-Turjman
Antalya Bilim University, Antalya, Turkey
e-mail: gozdegul.baser@antalya.edu.tr; oguz.dogan@antalya.edu.tr; fadi.alturjman@antalya.edu.tr

F. Al-Turjman (ed.), *Artificial Intelligence in IoT*,
Transactions on Computational Science and Computational Intelligence,
https://doi.org/10.1007/978-3-030-04110-6_4

to respond to new tourists' demands, and this response can be achieved by using innovative techniques offered by digitalization or new technology [6].

Tourists have different needs, expectations, and characteristics, and they prefer different types of tourism. Tourism destination covers interconnected services and tangible and intangible products and services. Smart tourism destination is a new but rapidly developing concept which needs considerable attention. This paper seeks to clarify the conceptualization of smart tourism destination and its link with the smart city concept as well as proposing a model for the smart city case study here in Antalya.

2 Smartness, Smart City, and Smart Tourism Destination

The notion of smartness finds its origin in the 1990s, although it proliferated significantly after 2008 [7]. Initially, the concept was coined as a complex technological infrastructure, embedded within urban areas to foster economic, social, and environmental prosperity [8]. More specifically, it posited the integration of ICT to improve processes and interconnect sub-systems [9] to ultimately tackle the economic, social, and environmental challenges imposed by urbanism [10]. Urbanism and managing big cities have enabled the smart city concept. A smart city is a city that uses advanced information and communication technology (ICT) to optimize resource production and consumption [11]. The concept of the smart city represents an environment where technology is embedded within the city. This technology will synergize with city's social components in order to improve citizens' quality of life, while also improving city services efficiencies such as optimizing the use of energy and better traffic monitoring [12]. The ultimate goal of smart places is to enhance the quality of life of all stakeholders, including residents and tourists [10].

Smartness is centered on a user perspective, which makes it more user-friendly than intelligent [1]. The term has been added to cities (smart city) to describe efforts aimed at using technologies innovatively to achieve resource optimization, effective and fair governance, sustainability, and quality of life [13]. The concept of the smart city represents an environment where technology is embedded within the city. This technology synergizes with city's social components in order to improve citizens' quality of life while also improving city services efficiencies, such as optimizing the use of energy and better traffic monitoring [1]. In connection with physical infrastructure (smart home, smart factory), the focus is on blurring the lines between the physical and the digital and on fostering technology integration. Added to technologies (smartphone, smart card, smart TV, etc.), it describes multi-functionality and high levels of connectivity. In addition, a smart city uses digital technologies to enhance performance and well-being, reduce costs and resource consumption, and engage more effectively and actively with its citizens.

Analyzing the concept of smart city, Cohen [14] defined the Smart City Wheel dimensions as (1) smart governance that relates with aspect of transparency with

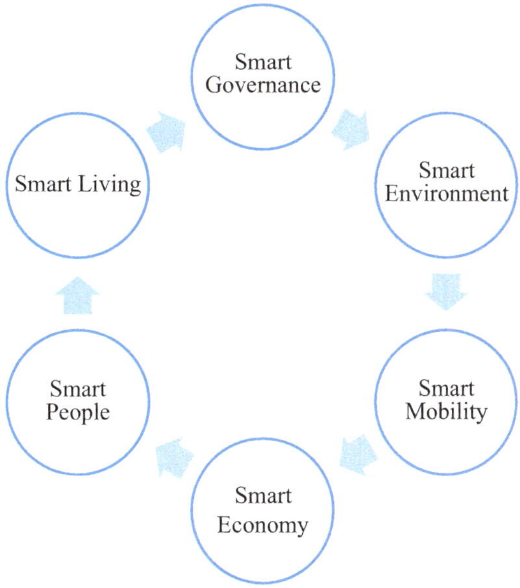

Fig. 1 [!b] Cohen's (2012) Smart City Wheel [14]

governance systems through modernization of city administration by supporting data openness and public involvement; (2) smart environment which is related to energy optimization that leads to sustainable management of available resources; (3) smart mobility which referred to accessibility within the city as well as outside the city and availability of modern transportation systems; (4) smart economy which is related to implementation of economic strategies based around digital technology; (5) smart people which linked to the qualification level of city's human capital; and (6) smart living which involves the quality of life which is measured in terms of healthy environment, social cohesion, tourist attraction, and availability of cultural and educational services (Fig. 1).

Smart city intertwines many entities to each other and therefore needs coordination, information, and infrastructure. Komninos et al. [15] indicate the base of the smart city as human capital, infrastructure/infostructure, and information. A well-structured smart city design is expected to support smart tourism destination as infrastructure and operation planning in the Internet of Things (IoT) era.

Smart city applications of the IoT involve large-scale deployments of wireless sensor networks (WSNs) which have gained lots of research attention in recent times. The SmartSantander project in Spain is an example of one such city-scale research project involving the deployment of over 3000 sensor and relay nodes within the city, supporting multiple applications [16]. Environmental monitoring, outdoor parking area management, and park and garden irrigation are some of the many use cases being tested on the IoT test-bed deployed in the city. The Santander project and other such smart city projects (F. Al-Turjman, et al., [17]) have provided a platform for researchers to experiment with the routing protocols, network coding schemes, and data mining techniques in a large-scale, multiuser application platform for sensor networks. The sensor network has to deal with large amounts of data, support requests from multiple users, and support information extraction from the network rather than serving as point-to-point communication network and transmit data from multiple information sources to the sink. Such a framework supports different types of users as the IoT user base, including individual users, private data centers, and government agencies. These users interact with the sensor network through an Internet-based interface access network, which is perceived to be an information-centric network (ICN) in the future [18–22]. Data gathered from the sensor network is delivered to the interface access network through gateway nodes. These gateway nodes could either communicate directly with the sink node of the sensor network or could be distributed throughout the network to provide multiple access points. Currently, the experiments are focused on enabling each application as a separate entity. However, in the real world, many of the applications run simultaneously, and the network receives requests from multiple end users in IoT user base at the same time. For example, in a smart city application, the same infrastructure (sensor and relay nodes) that is used in periodic monitoring of traffic intensity is also used for outdoor parking area management and to provide information about traffic congested areas to users on demand. It can even be used to send out high-priority alerts about hazardous road conditions or accidents to users. The information generated by each of these request types has different attributes associated with it. While the periodic monitoring information needs to be reliable, it does not have a strict upper bound on the time taken to gather and deliver the data, as long as it happens before the end of the stipulated time period. However, for on-demand requests generated by the user, such as a user requesting to know the availability of a free parking space in a region, the information has to be delivered quickly (low latency) to the user. In case of emergency alerts, the information must be transmitted reliably and as quickly as possible, to all users in the area. This shows that the sensor network must be able to segregate the requests and manage the heterogeneous traffic flows in a way that satisfies the end user in terms of the perceived quality of information for each request type. Latency, reliability, accuracy, relevance, and robustness are some of the attributes that collectively provide an estimate of the quality of information (QoI) perceived by the user [18–22]. To enable such QoI-aware data delivery, the use of artificial intelligence (AI)/cognition approaches in the underlying sensor network is recommended [23–25] – [18–22]. Cognition refers to the ability to be aware of the environment, in addition to be able to learn from the past actions and use it in making future decisions that benefit the network [23–25] – [18–22]).

Smart tourism destination initiates back to the developments in technology starting with digitalization. Digitalization in tourism has initiated since the 1950s as being parallel to the advances in technology. Broadly speaking, digital tourism is concerned with the use of digital technologies to enhance the tourist experience before, during, and after the tourist activity [26]. Lopez de Avila [27] defines smart tourism destination as an innovative tourist destination, built on an infrastructure of state-of-the-art technology guaranteeing the sustainable development of tourist areas, accessible to everyone, which facilitates the visitor's interaction with and integration into his or her surroundings, increasing the quality of the experience at the destination and improves residents' quality of life.

Digital tourism focuses on a wide variety of tourist activities in destinations, e.g., museums, transportation systems, concerts, activities, rallies, countrysides, zoos, theme parks, etc. [4]. Tourism apps, events information, accommodation reservation platforms, tourist card, online tickets, integration of cards, and platforms with other services are some components of smart tourism initiatives. Younger tourists and tourists with higher incomes tend to use e-services more intensively [28]. In this regard, digitalization, namely, e-tourism, provides some advantages like the reduction of seasonality, the more successful communication with the customers, and the rise in reservations and sales in general ([29], p. 336).

In addition to the rapid developments in digitalization, tourism services still experience very relevant changes [30], due to the unprecedented development of information and communication technology (ICT) in recent decades [28]. ICTs create value-added experiences for tourists [1] and new opportunities for marketing and distribution strategies for tourism providers [28]. From the supply of products to information search process and consumption patterns, tourism experiences and their preparations can be progressively transformed by advances in ICT [31]. In regard to smart tourism destination, smart city concept is required to be considered as many applications either coincide or complete each other. Both smart tourism destination and smart city concepts aim to achieve a better service for tourists and local people by providing fast, reliable, and rich information and practices.

Basing on the smart city concept, smart tourism destination contributes to the value creation through collecting, sharing, and organizing data supported by the smart city infrastructure. This infrastructure resembles an ecosystem in which stakeholders interact with each other continuously. Therefore, the ecosystem approach has been recognized as suitable to address the topics of smart cities and smart tourism destinations methodologically [11, 13, 32]. The ecosystem in the smart tourism destination provides tourists with real-time and personal services and simultaneously collects data for the optimization of their strategic and operational management [11, 13, 33]. From the smart destination (SD) logic perspective, an ecosystem has been outlined as a relatively self-contained, self-adjusting system of resource-integrating actors connected through shared institutional logics and mutual value creation through voluntary service exchange ([34], p. 15). In this respect, the tourist can collect information, before, during, and after the visit [35]. Also, developments in geographical information systems (GIS) have created new opportunities for tourism marketing and promotion [36]. Smart tourism destinations take advantage of (1) technology-embedded environments, (2) responsive processes at micro and macro levels, (3) end-user devices in multiple touch points, and (4) engaged stakeholders that use the platform dynamically as a neural system ([1], p. 557).

Smart cities can be stated as the seed of smart destinations. However, there are important differences between the terms:

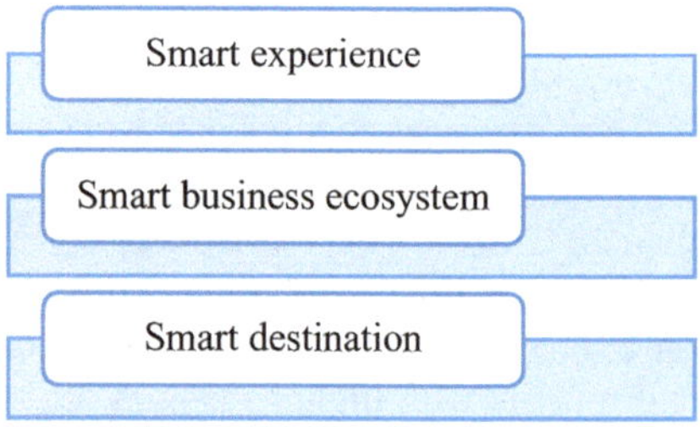

Fig. 2 Smart tourism components

- The main target audience of a smart destination is the tourist not so much the locals. In essence, smart cities and smart destinations share infrastructure as well as facilities while providing solutions to locals and tourists. Therefore, multilingualism, cultural differences, culinary uses, the seasonality of the visiting population, etc. should be taken into consideration [37].
- A smart destination is driven by the tourism sector and public as well as private institutions. Its governance is shared through the formation of inclusive entities such as boards, trusts, foundations, etc.
- Smart destinations are bound to the increase in their competitiveness and to the improvement of the tourist experience.

Smart tourism has three basic components such as smart experience, smart business ecosystem, and smart destination. The smart experience component specifically focuses on technology-mediated tourism experiences and their enhancement through personalization, context awareness, and real-time monitoring Buhalis and Amaranggana (2014). In a smart tourism ecosystem, any type of stakeholder can become a producer, consumer, intermediary, etc. depending on resources and connections rather than predefined roles [11, 13]. Enabling tourism destination managers to understand the importance of integrating smartness for value co-creation can enhance competitiveness ([5], p. 109) (Fig. 2).

Furthermore, smart tourism destinations can be structured as having six A's as (1) attractions which can be natural such as mountain, artificial such as amusement parks, or cultural such as a music festival; (2) accessibility which refers to the entire transportation system within the destination that comprises of available routes, existing terminals, and adequate public transportations; (3) amenities which characterize all services facilitating a convenient stay, namely, accommodation, gastronomy, and leisure activities; (4) available packages which refer to the availability of service bundles by intermediaries to direct tourists' attention to certain unique features of a respective destination; (5) activities which refer to all available activities at the destination which mainly trigger tourists to visit the destination; and (6) ancillary services which are those daily use services which are not primarily aimed for tourists such as bank, postal service, and hospital [38]. Destinations need to interconnect all their stakeholders to facilitate a dynamic co-creation process to increase destination competitiveness [39]. Smart tourism destination may cover the

Fig. 3 6 A's of smart tourism destination

following applications appropriate for the six A's before, during, and after the visit of the tourist (Fig. 3).

3 Smart Tourism Destination Instruments and Platforms

The instruments and platforms required to develop a smart tourist destination include the following: [6]

1. *ICT*: The key aspect of smart destinations is the integration of ICTs into physical infrastructure [1]. ICT could contribute in terms of generating value-added experiences for tourists, while also improving efficiency and supporting process automation for the related organizations (11, 13], p. 180). One of the basic changes of digital tourism is the opportunity it creates for tourists to design their own tours and seek out destinations, hotels, flights, etc. upon their individual preferences through Internet access. According to the "World Travel Trends – 2016" report, the Internet is far and away the best tool in travel searching [40]. Tourists browse the Internet to collect traveling information and visit related tourism destination's website which becomes a very useful tool to promote tourism by giving important information for visitors [4]. With 60% of leisure and 41% of business travelers making their travel arrangements via the Internet [41], the Internet has become one of the most important communication tools for tourists as well as tourism providers [42]. Da Costa Liberato et al. [6] claim that

travel trends over the next few years will be determined by the intensive use of the Internet.

2. *Cloud computing*: The use of cloud computing could reduce fixed costs and shift them into variable costs based on the necessities. It also stimulates information sharing that is fundamental to undertake smart destinations.
3. *Internet of Things (IoT)*: IoT is a network that connects anything in anytime and anyplace in order to identify, locate, manage, and monitor smart objects [43]. The recent technological development, however, has enabled the rise of Internet of Things paradigm, in which devices are connected to the Internet, currently allowing 1.6 billion people to have constant access to information [44]. The IoT is a combination of people, processes, data, and everything that makes networks more relevant and valuable than ever before, turning information into actions that create new resources, richer experiences, and unprecedented economic opportunities for companies, individuals, and countries [6]. The IoT could support smart destinations in terms of providing information and analysis as well as automation and control. The process of applying IoT to tourism has to undergo several stages, namely, tourism infrastructure construction, tourism information data construction, and tourism service platform construction. Ultimately, the unified information platform, which will support tourism operation monitoring and automated management, is created [45]. The basic idea of the IoT is the pervasive presence around us of a variety of objects such as radio-frequency identification (RFID) tags, sensors, actuators, mobile devices, etc. which are able to interact with each other and cooperate with their neighboring objects to achieve common goals [46].
4. *End-user Internet service systems*: It refers to a number of applicants at various levels supported by a combination of cloud computing and IoT [6]. End-user Internet service systems include the websites and social media platforms. The websites of tourism destinations grant detailed information about tourism products and services of the location, reduction in dependence to agents, ability to compare alternative products and services during decision-making phase, making reservations directly, and buying the product or service individually [47]. The well-known and most visited destinations have an official website. Table 1 shows the official websites of some of the destinations. These official websites provide some information about services provided in the destinations such as café and restaurants, accommodation, transportation, events, nightlife, what to visit, etc.

Secondly, social media plays an important role in decision-making process [48]. A research conducted with travelers from the USA, Australia, the UK, and Canada implies that 36% of tourists use social media and blog sites, while 54% of them use travel review sites [48]. In 2017, 58.2% of American travelers used user-generated content, and 54.6% used social media in the travel planning process. Fifty-two percent of Facebook users said friends' photos affected their travel plans [49]. Besides Facebook, TripAdvisor, YouTube, Twitter, and LinkedIn are also used by tourists [40]. Carter [49] asserts that 29% of travelers use Facebook, 14%

Table 1 Official websites of destinations

Destinations	Websites
Barcelona	http://www.barcelonaturisme.com/
Madrid	http://www.esmadrid.com/
Stockholm	http://www.visitstockholm.com/
Rome	http://www.turismoroma.it/
Sydney	https://www.sydney.com/
London	https://www.visitlondon.com/
Paris	https://en.parisinfo.com/
Berlin	https://www.visitberlin.de/en

use TripAdvisor, and 6% use Twitter before the travel planning. Social media is used not only before vacation but also during and post vacation. Especially before and during the trip, they rely on mobile technologies to simplify the travel by searching for information about transportation, accommodation, attractions, and activities [50]. Seventy-two percent of people share post-vacation photos on social media, and 70% update their Facebook status while they are still on vacation. In the post-vacation process, 76% post their vacation photos on social media, 55% like Facebook pages specific to a vacation, and 46% post hotel reviews [49]. As a result, tourism providers concern social media significantly through different channels. Mistilis and Gretzel [51] stress that 50% of Australian tourism providers use social media.

5. *Mobile apps*: Tourists get access to end-user Internet service systems through their mobile devices such as smartphones and tablets besides their computers or the available computers they may use. The accessibility of easy-to-carry devices like these allows for the easy purchase of goods and services via mobile devices instead of through more conventional means [6]. Due to the widespread use of the Internet, mobile devices have also become a phenomenon [52]. Furthermore, smartphones have brought many changes to the tourism industry [53]. Multiple studies showed that an increasing number of consumers are booking travel arrangements on their mobile phones – a number that has been increasing steadily for the past six quarters. The travel industry, specifically, boasts over 50% growth year-over-year in mobile [54]. In the USA, almost 40% of tourists access travel information, while 25% book their trips using smartphones [53]. Sorrels [55] claims that 37% of consumers make shopping for flights and 43% make a booking via smartphones. In 2017, 58.9% of American travelers used a mobile phone to access travel info. The digital travel sales are expected to reach 140 billion dollars by 2018 [56]. Mobile apps provide several benefits to tourist. First, some websites are not suitable for mobile using, and mobile apps can overcome this problem. The menus of smartphone applications contain panoramas, videos, interpretation manuscripts, background music, narration, games, AR-based path finding, and SNS connection services (e.g., Facebook, Twitter, KakaoTalk) [57]. Additionally, it can be personalized for each tourist. Secondly, tourists can access restaurants, transportations, local activities, etc. using it. Mobile apps allow the

tourist to reach information [58] and to purchase products and services [53] without the need of a web browser. Therefore, tourists can plan their travels or make room reservations through mobile apps [6]. According to TripAdvisor, 60% of smartphone users have downloaded travel apps, and 45% plan to use the apps for travel planning. Furthermore, 55% of travel apps are purchased within 3 days of travel or while travelers are at the destination. It shows the importance of mobile apps in the tourism industry. Thus, some destinations have launched their mobile applications to customers [53]. Singapore, Korea, London, and Brisbane are among the destinations that offer mobile applications ([11, 13, 45, 58, 59]). Moreover, some hotel chains, luxury resorts, and 5-star hotels have also launched their mobile applications to customers [50, 53]. Tourists could make a room reservation and access the guest loyalty program using the mobile app [53].

6. *Mobile payment*: Smart tourism is not only about being on the Internet or using mobile apps but also using other digital resources such as mobile payment that is receiving growing interest globally as an alternative to using cash, check, or credit cards [60]. Peng et al. [61] describe the mobile payment the use of a mobile device to conduct a tourism payment transaction in which money or funds are transferred from a payer to a receiver via an intermediary or directly, without an intermediary in the tourist destination. The tourists make mobile payments using mobile wallet which comprises of mobile phone with consumer-specific information that allows guest to make payment [6] via various wireless technologies [60] such as near field communication (NFC), radio-frequency identification (RFID), and unstructured supplementary service data (USSD) [6]. Mistilis and Gretzel's [51] study shows that 70% of tourism providers operating in Australia offered online/mobile payment facilities.
7. *Virtual/augmented reality*: It is a computer-generated three-dimensional environment that allows users to view, interact, and manifest their presence in a non-real environment. Depending on the interactivity provided, it may be immersive (e.g., using a helmet) or non-immersive (e.g., using monitors). Given the great potential of this technology in many areas, especially in tourism, there has been a widespread proliferation of mobile applications (apps) with augmented reality. These are used in museums, monuments, galleries, open spaces, and other tourist attractions where objects can be enhanced and supplemented, in real time, with a variety of information (text, images, three-dimensional animation, audio, or video) [6]. Augmented reality (AR) can be described as a combination of technologies that enable real-time mixing of computer-generated content with live video display. In other words, AR allows the physical world to be enriched digitally. AR augments user's view and transforms it with the technologic devices such as a computer, mobile phones, etc. [62]. Thus, the user can view the real world augmented with additional 3D graphics layered to his/her field of vision [63]. So, AR increases the user's reality perception and surrounding environment perception [62]. Lancaster, Vienna, Basel, Belgium, and London are some of the examples to mobile tourist guide applications [62, 64, 65].
8. *Artificial intelligence (AI) and cognition in smart cities*: Various artificial intelligence (AI) techniques have been applied to WSNs in smart cities to improve

their performance and achieve specific goals. We look at AI techniques as a means of introducing learning in the WSN. Learning is an important element in the observe, analyze, decide, and act (OADA) cognition loop [12, 66], used to implement the idea of cognitive wireless networks [67]. In fact, we can classify AI techniques into computational intelligence (CI) techniques, reinforcement learning (RL) techniques, cognitive sensor networks, multi-agent systems (MAS), and context-aware computing. Although these techniques are closely related with each other, we segregate them to show the different goals that learning can achieve for the network as follows.

3.1 Computational Intelligence

CI techniques are a set of nature-inspired computation methodologies that help in solving complex problems that are usually difficult to fully formulate using simple mathematical models. Examples of CI techniques include genetic algorithms, neural networks, fuzzy logic, simulated annealing, artificial immune systems, swarm intelligence, and evolutionary computation. In a learning environment, CI techniques are useful when the learning agent cannot accurately sense the state of its environment. However, a major drawback of this methodology is that it can be computationally intense and may require some form of model-based off-line learning to deliver to the requirements of the application scenario. Techniques such as ant colony optimization, for example, can cause an undesirable increase in communication overhead in the smart city WSNs too [23–25].

3.2 Machine Learning

Machine learning can be classified into supervised, unsupervised, and reinforcement learning. Supervised learning would be more compute intensive and requires a training sequence. Additionally, accuracy of the learning algorithm would then be defined by this training sequence. In the unsupervised learning approach, the learning is from the environment being observed, and no training sequence is required. Reinforcement learning (RL) is a reward-based technique that emphasizes on learning while interacting with the environment, without relying on explicit supervision or complete model of the environment. It is a method of automating goal-directed learning and decision-making. In smart cities' WSNs, RL has been successfully applied in networking tasks such as adaptive routing, identifying low cost and energy-balanced data delivery paths [68], and in information processing tasks involving data aggregation and inference [69].

3.3 Cognitive Networks and Multi-agent Systems

Cognitive networks are built around the idea of having sensor networks evolve around user requirements. It is about taking a step toward developing intelligent networks that do not limit themselves to point-to-point communication within the network. Instead, they enable the network to perceive user requirements and deliver data using distributed intelligence in the network. To implement distributed intelligence in smart cities, the multi-agent systems (MAS) are typically used. The agents in these MAS are called cognitive agents. They may interact to achieve information fusion and retrieval and may also be able to predict data for future use.

3.4 Context-Aware Computing

In large-scale WSNs such as those in smart cities' applications, a huge amount of data is generated. In order to derive useful information from raw data, context of the data plays an important part. Context awareness is even more important in the IoT era, as it enables the network to deliver relevant, user-requested data. While doing so, network resources are also conserved by extracting only meaningful information that is relevant to the requests, from the network. There are various aspects to context-aware computing. They are context acquiring, context modeling, context reasoning, and context distribution [70]. Context awareness is very important and valuable in IoT-based smart cities' applications, as it can add value to the large amount of data available from these applications [23–25].

4 The STD Framework

The case study approach is regarded to be appropriate in this study since it includes the analysis of reports, studies, news, articles, and other text-sensitive documentation and provides a comprehensive coverage of information. According to Simon et al. [71], the case study is a focusing approach that allows the discovery of a variety of interactive processes and the factors involved in an in-depth study of a destination. It is a flexible process, taking into account unexpected issues that may arise or which a participant deems important.

The case study as a research method can include various techniques. Among primary research, the most frequently used methods are observation and interviews [71]. The case study method is, therefore, appropriate to achieve the objectives of the study since it includes the analysis of reports, studies, news, articles, and other text-sensitive documentation and provides a comprehensive coverage of information. In-depth interviews are also made with officers who are in charge of promotion of Antalya (Ministry of Culture and Tourism) and smart city project (Antalya Municipality).

The study consists of three stages:

1. A comprehensive literature review of smart technology which could be applied for tourism
2. Determining the current applications in Antalya tourism destination
3. Proposing a model for Antalya to be a smart tourism destination

4.1 Antalya as an STD

Antalya, an important destination of the Turkish tourism industry, is located in the Southern Mediterranean of Turkey. Antalya has been very popular as tourism destination because of its clean beaches, sea, Mediterranean climate, and high number of historical and natural sights. In addition, one of the main characteristics of the region is having new, high quality, and a high number of accommodation facilities compared to competitor countries and regions. As of 2018, Antalya as a tourism destination has 590.000-bed capacity [72].

Smart tourism destination is a new concept for Antalya, and it is mainly attached to "smart city" applications. In the first months of 2015, the Metropolitan Municipality of Antalya announced that it started a project to become a smart city in tourism, transportation, health, security, and municipal services. In addition, it was announced that free Internet, electronic traffic control system, smart public transportation system, intelligent tourism points, intelligent health service, unobstructed SMS system, intelligent environment, and smart energy systems would be established within the scope of the project ([73], p. 158). The current state of Antalya in the context of smart tourism can be summarized as follows in Table 2.

4.2 An STD Model for Antalya

The smart tourism process can be examined in three phases, namely, before the vacation, during the vacation, and post-vacation. In before-the-vacation phase, tourists try to get information about the destination and try to plan their vacation. Mostly they start by visiting the website of the planned destination and the travel blog sites. In this respect, smartphone applications can also be considered as an efficient tool because it enables tourists to find information anywhere and anytime in accordance with where the users are and what situation is through the latest technology, including applications, augmented reality (AG), and location-based service (LBS) [58]. The social media can also affect the tourists' decision where a destination to be visited has not yet been decided at this stage. The attractive or interesting photos about a destination shared by the service providers or tourists'

Table 2 A summary of Antalya city STD

	Website	App	AR	Virtual tour
Antalya	www.antalyadestination.com www.visitantalya.com www.visit-antalya.com www.antalyatourguide.org www.antalyakulturturizm.gov.tr www.antalya.gov.tr www.antalya.bel.tr www.antalyamiz.com www.antalyatourguide.com	Antalya Travel Guide and Offline City Street Map/Antalyakart/ ICF Airport mobile app	Antalya Travel Guide and Offline City Street Map	antalyacentral.com/ Ministry of Culture and Tourism/ mekan360.com
Kemer	www.visitkemer.com	Kemer Travel Guide		
Manavgat	www.visitmanavgatside.com	–		
Side	www.visitmanavgatside.com	Antalya Travel Guide and Offline City Street Map		Ministry of Culture and Tourism
Belek	www.visitbelek.com	–		
Serik		Serik Tourism Guide		
Demre	hometurkey.com	–		Ministry of Culture and Tourism

friends on social media can attract tourists' attention to the destination. When the destination to be visited is decided, tourists search available hotels operating in the destination using travel blogs or social media, and finally, they make hotel booking via the hotel or travel agency websites. Tourists can also make hotel booking via hotels' social media accounts. Due to the technologic development, nowadays some hotel and travel companies provide booking service via their mobile apps. Thus, tourists can prefer hotel and travel booking using their mobile apps.

In during-the-vacation phase, tourists are expected to have a high level of satisfaction from their vacations; this could be supported by supplying them easy access to information as well as high-quality and fast information. This section covers the transportation, sightseeing, accommodation, dining, shopping, and all other facilities demanded by a tourist. As an example, some hotels allow tourists keyless room entering via mobile apps. Free applications provide information about the museums, restaurants, café-bar, etc. located at the destination. Thus, tourists can make a tour plan or make a sightseeing using the mobile apps. During the sightseeing, they can take photos and can share them on their social media accounts by mentioning the destination or the hotel.

In the post-vacation phase, tourists can share photos related to their vacations. They can also share thoughts about the destination or hotel on travel blogs. Post-vacation stage is crucial since the tourists may decide to revisit and share their impressions for the destination. Table 3 shows the model for Antalya as a smart tourism destination.

Table 3 A model for Antalya as a STD

	Before vacation	During vacation	Post-vacation
Smart tourism office	Antalya website (1)		
Hotel management		Digital accommodation (2)	
Smart tourism office		Hotel receptions (3)	
Smart tourism office			Follow-up messages (4)
Smart tourism office			
Smart tourism office		Mobile app for Antalya (5)	
Antalya Municipality		Antalya transportation app for tourists (6)	
Ministry of Culture and Tourism		Digital tourist kiosks (7)	
Smart tourism office	IoT (8)	IoT	IoT
Antalya Municipality		Free Wi-Fi hotspots (9)	
Smart tourism office	Social media (10)	Social media	Social media
Smart tourism office		QR codes (11)	
Smart tourism office		Virtual reality (12)	
Smart tourism office	Artificial intelligence (13)	Artificial intelligence	Artificial intelligence
Smart tourism office		Smart tourism kiosks (14)	
Smart tourism office	Design your stay in Antalya (15)		Share your stay in Antalya (16)
Smart tourism office	Social media (17)	Social media (17)	Social media (17)
Smart tourism Office	Bloggers (18)	Bloggers (18)	Bloggers (18)

1. Antalya website: websites for Kemer, Side, Manavgat, Alanya, Kundu, etc. combined with Antalya website
2. Digital accommodation: hotels offering digital services like hotel apps and mobile payment
3. Augmented reality: virtual augmented reality corners for tourists to see a short vision for historical sites
4. Follow-up messages: asking for a rating for vacation and asking the intend to visit again by a message to phone or e-mail
5. Mobile app for Antalya: instant mobile app access for every tourist arriving at Antalya airport, covering information related to transportation, climate, sites, etc.
6. Antalya transportation app for tourists in different languages
7. Digital tourist boxes: offices offering information, augmented reality, and all kinds of tourist information
8. IoT: Shopping, dining, social media, advertisement, etc.
9. More free Wi-Fi hotspots in Antalya
10. Social media accounts of the smart tourism office in Antalya
11. QR codes in several historical, natural sites, city center, etc.
12. Virtual reality for historical sites
13. Artificial intelligence
14. Smart tourism kiosks
15–16. Websites for Antalya
17. Social media accounts
18. Bloggers for Antalya

5 Concluding Remarks

The study proposes a model for Antalya as a smart tourism destination. The linkage between one tourism product and another at destination level is vital because of the nature of tourism industry which are combinations of multiple components served in several touch points that are perceived by the customer prior, during, and after their trip [74]. Therefore, the model needs to be organized, coordinated, and controlled by a "smart tourism office" to be established by a group of experts from tourism sector, university, engineers, municipality, and government.

The model focuses on attempts to be made before, during, and after vacation for a tourist. The main goal is to provide a higher-quality tourist experience by providing a fast, easy, and high-quality access to services, tourist attractions, gastronomy, accommodation, transportation, etc. For this purpose, in the "before vacation" stage, tourists need proper information through websites, apps, IoT, social media, design your vacation site, and bloggers. For the time being, Antalya has many websites which may cause a confusion for potential tourists. During vacation stage, tourists need to plan their stay as well as having a good organization, joy, and satisfaction which can be achieved by digital accommodations, apps, smart transportation, kiosks, QR codes, virtual reality, IoT, and artificial intelligence. After vacation stage requires follow-up messages, IoT, and social media.

The main benefit of the model would be to provide tourists to get fast, reliable, easy information which would increase the quality of their tourism experience. Today, tourism industry is subject to the technological transformation that can make the conducting of business easier and faster and the transmission of information more convenient [6]. The model would enable tourists access to services, touristic attractions, shops, transportation, local food, hospitals, etc. easily and safely.

Secondly, smart tourism applications like augmented reality and apps may increase the positive impression of the tourists for the tourism destination as they may get interesting information which may create excitement and pleasure. Next, they may help other travelers in their decision-making process, revive and reinforce their travel experiences, as well as construct their self-image and status on social networks ([11, 13], p. 181). This would increase the destination's image and would attract more tourists to the destination.

Tourism revenue, has a contribution of 3.1% to Turkish GDP in 2017, which influences its impact and importance for the economy. There seems to be many applications and practices of ICT to be adapted for tourism. Tourism industry can be one of the driving forces of modern economies as it is a leader user of ICT [6]. Tourism industry is expected to grow vastly in Turkey and influence the ICT sector positively.

Finally, there needs to be a council, board, or an office to coordinate the smart tourism destination network which would follow all kinds of steps, applications, practices, etc. Becoming a smart tourism destination requires leadership, vision, patience, strategic management, and continuous evaluation and change. Perceiving the smart tourism destination as an ecosystem is essential, and a vision and a clear set of goals for innovation are key facilitators for developing smart tourism destinations as a collective whole ([5], p. 119). In addition, smart tourism applications require a continuous follow-up for 24 h. For this purpose, the smart tourism destination council can be formed by a group of engineers, tourism experts, municipality officers, etc. Public-private partnership (PPP) is essential when running a smart tourism destination initiative. Enhancing collective intelligence is essential in smart tourism destinations ([5], p. 118).

Another fact to consider is the "smart city" and smart tourism destination link. Smart city infrastructure and applications interact with smart tourism destinations. Therefore, smart tourism destination would promote and improve fastly in case of smart city practices. Smart tourism destinations should also perform smartness by implementing appropriate tourism applications within smart cities' components as defined by [14]. To take full advantage of the current possibilities provided by smartness, destination managers have to integrate the entire range of smartness components and ensure interoperability and interconnectivity of both soft and hard smartness ([5], p. 120). Also, smart tourism destinations need to emphasize ways to enhance the tourist experience, while simultaneously improving the quality of life for residents.

Local and foreign tourists need information about how to use and access to smart tourism destination network and content. It is recommended that destinations not only focus on exploiting the use of new technology but also educate the citizens and visitors on how to best use this new technology.

Finally, to survive, traditional tourism firms have to redefine their business model and the way they propose to create customer value ([11, 13], p. 183). Smart tourism destination can be an important tool for creating customer value by taking the advantage of technology.

References

1. Buhalis, D., & Amaranggana, A. (2014). Smart tourism destinations. In *Information and Communication Technologies in Tourism 2014*. Proceedings of the International Coonference in Dublin Ireland, January, 21–24, (pp. 553–564).
2. Sadr, S. M. H. (2013). The role of ICT in tourism industry on economic growth: Case study of Iran. *European Journal of Business and Management, 5*(17), 159–165.
3. Wahab, S., & Cooper, C. (Eds.). (2001). *Tourism in the age of globalisation*. Routledge.
4. Putra, F. K. K., Saepudin, P., Adriansyah, E., & Adrian, I. G. A. W. (2018). Digital tourism: A content analysis of West Java tourism websites. *Journal of Indonesian Tourism and Development Studies, 6*(2), 73–84.
5. Boes, K., Buhalis, D., & Inversini, A. (2016). Smart tourism destinations: Ecosystems for tourism destination competitiveness. *International Journal of Tourism Cities., 2*, 108. https://doi.org/10.1108/IJTC-12-2015-0032.
6. da Costa Liberato, P. M., Alén-González, E., & de Azevedo Liberato, D. F. V. (2018). Digital technology in a smart tourist destination: The case of porto. *Journal of Urban Technology, 25*(1), 75–97.

7. Hollands, R. G. (2015). Critical interventions into the corporate smart city. *Cambridge Journal of Regions, Economy and Society, 8*(1), 61–77.
8. Meijer, A., & Bolívar, M. P. R. (2016). Governing the smart city: A review of the literature on smart urban governance. *International Review of Administrative Sciences, 82*(2), 392–408.
9. Townsend, A. M. (2013). *Smart cities: Big data, civic hackers, and the quest for a new utopia.* WW Norton & Company.
10. Caragliu, A., Del Bo, C., & Nijkamp, P. (2011). Smart cities in Europe. *Journal of urban technology, 18*(2), 65–82.
11. Gretzel, U., Werthner, H., Koo, C., & Lamsfus, C. (2015). Conceptual foundations for understanding smart tourism ecosystems. *Computers in Human Behavior, 50*, 558–563.
12. Vicini, S., Bellini, S., & Sanna, A. (2012, May). How to co-create internet of things-enabled services for smarter cities. In *The First International Conference on Smart Systems, Devices and Technologies* (pp. 55–61).
13. Gretzel, U., Sigala, M., Xiang, Z., & Koo, C. (2015). Smart tourism: Foundations and developments. *Electronic Markets, 25*, 179. https://doi.org/10.1007/s12525-015-0196-8.
14. Cohen, B. (2012). Smart cities hub. http://smartcitieshub.com/2012/11/11/smart-cities-ranking-methodology/
15. Komninos, N. (2013). *Intelligent cities: innovation, knowledge systems and digital spaces.* Routledge.
16. SmartSantander, (2018) Future Internet Research and Experimentation. [Online]. Available: http://www.smartsantander.eu/.
17. Al-Turjman, F., Ever, E., & Zahmatkesh, H. (2018). Small cells in the forthcoming 5G/IoT: traffic modelling and deployment overview. *IEEE Communication Surveys and Tutorials.* https://doi.org/10.1109/COMST.2018.2864779.
18. Al-Turjman, F. (2017). Information-centric framework for the internet of things (IoT): traffic Modelling & Optimization. *Elsevier Future Generation Computer Systems, 80*(1), 63–75.
19. Alabady, S., & Al-Turjman, F. (2018). Low complexity parity check code for futuristic wireless networks applications. *IEEE Access Journal, 6*(1), 18398–18407.
20. Al-Turjman, F. (2017). Mobile couriers' selection for the smart-grid in smart cities' pervasive sensing. *Elsevier Future Generation Computer Systems, 82*(1), 327–341.
21. Al-Turjman, F. (2017). Energy–aware data delivery framework for safety-oriented mobile IoT. *IEEE Sensors Journal, 18*(1), 470–478.
22. Al-Turjman, F. (2017). A cognitive routing protocol for bio-inspired networking in the Internet of Nano-Things (IoNT). *Springer Mobile Networks and Applications.* https://doi.org/10.1007/s11036-017-0940-8.
23. Al-Turjman, F. (2018). QoS–aware data delivery framework for safety-inspired multimedia in integrated vehicular-IoT. *Elsevier Computer Communications Journal, 121*, 33–43.
24. Al-Turjman, F. (2018). Fog-based caching in software-defined information-centric networks. *Elsevier Computers & Electrical Engineering Journal, 69*(1), 54–67.
25. Al-Turjman, F. (2018). Modelling Green Femtocells in Smart-grids. *Springer Mobile Networks and Applications, 23*(4), 940–955.
26. Benyon, D., Quigley, A., O'Keefe, B., & Riva, G. (2014). The presence and digital tourism. *AI & SOCIETY, 29*, 521–529.
27. Lopez de Avila, A. (2015, February). Smart destinations: XXI century tourism. In *ENTER2015 conference on information and communication technologies in tourism*, Lugano, Switzerland.
28. Neuts, B., Romao, J., Nijkamp, P., & van Leeuwen, E. (2013). Digital destinations in the tourist sector: A path model for the impact of e-services on tourist expenditures in Amsterdam. *Letters in Spatial and Resource Sciences, 6*(2), 71–80.
29. Pitoska, E. (2013). E-tourism: The use of internet and information and communication technologies in tourism: The case of hotel units in peripheral areas.
30. Baggio, R., & Del Chiappa, G. (2014). Real and virtual relationships in tourism digital ecosystems. *Information Technology and Tourism., 14*, 3. https://doi.org/10.1007/s40558-013-0001-5.

31. Jacobsen, J. K. S., & Munar, A. M. (2012). Tourist information search and destination choice in a digital age. *Tourism Management Perspectives, 1*, 39–47.
32. Baron, G. (2013). Smartness' from the bottom up: A few insights into the Amsterdam smart city programme. *Metering International, 3*, 98–101.
33. Wang, X., Li, X. R., Zhen, F., & Zhang, J. H. (2016). How smart is your tourist attraction?: Measuring tourist preferences of smart tourism attractions via a FCEM-AHP and IPA approach. *Tourism Management*. https://doi.org/10.1016/j.tourman.2015.12.003.
34. Wieland, H., Polese, F., Vargo, S. L., & Lusch, R. F. (2012). Toward a service (eco) systems perspective on value creation. *International Journal of Service Science, Management, Engineering, and Technology (IJSSMET), 3*(3), 12–25.
35. Mansson, M. (2011). Mediatized tourism. *Annals of Tourism Research, 38*(4), 1634–1652.
36. Chang, G., & Caneday, L. (2011). Web-based GIS in tourism information search: Perceptions, tasks, and trip attributes. *Tourism Management, 32*(6), 1435–1437.
37. Khan, A. A., Rehmani, M. H., & Reisslein, M. (2017). Requirements, design challenges, and review of routing and MAC protocols for CR-based smart grid systems. *IEEE Communications Magazine, 55*(5), 206–215.
38. Buhalis, D. (2000). Marketing the competitive destination of the future. *Tourism Management, 21*(1), 97–116.
39. Neuhofer, B., Buhalis, D., & Ladkin, A. (2012). Conceptualising technology enhanced destination experiences. *Journal of Destination Marketing & Management, 1*(1–2), 36–46.
40. Acar, E. (2017). Turizmde yeni trendler-dijital turizm ve inovasyon. In III. Turizm Şurası (pp. 116–126).
41. Li, H., & Suomi, R. (2008). Internet adoption in tourism industry in China. In *Towards sustainable society on ubiquitous networks* (pp. 197–208). Boston: Springer.
42. Gelter, H. (2017). *Digital tourism – An analysis of digital trends in tourism and customer digital mobile behaviour*.
43. Mingjun, W., Zhen, Y., Wei, Z., Xishang, D., Xiaofei, Y., Chenggang, S., Xuhong L, Fang W, & Jinghai, H. (2012, October). A research on experimental system for Internet of things major and application project. In *System Science, Engineering Design and Manufacturing Informatization (ICSEM), 2012 3rd International Conference on* (Vol. 1, pp. 261–263). IEEE.
44. Fuentetaja, I. G., Simon, I. Z., Aranzabal, A. R., Ariza, M. P., Lamsfus, C., & Alzua-Sorzabal, A. (2013). An analysis of mobile applications classification related to tourism destinations. In *Information and Communication Technologies in Tourism 2014* (pp. 31–44). Cham: Springer.
45. Guo, Y., Liu, H., & Chai, Y. (2014). The embedding convergence of smart cities and tourism internet of things in China: An advance perspective. *Advances in Hospitality and Tourism Research (AHTR), 2*(1), 54–69.
46. Want, R., Schilit, B. N., & Jenson, S. (2015). Enabling the internet of things. *Computer, 48*(1), 28–35.
47. Köker, N. E., & Göztaş, A. (2010). Digitalization of the cities: An analysis of city municipality web sites as a part of city brand. *Journal of Yasar University, 20*(5), 3331–3347.
48. Expedia. (2017). Multi-National Travel Trends Connecting the Digital Dots: The Motivations and Mindset of Online Travelers. Retrieved from http://www.societyofpatriots.com/assets/files/Research/Multi-National_Travel_Trends_2017.pdf.
49. Carter, E. (2017). Social Media, Mobile, and Travel: Like, Tweet, and Share Your Way Across the Globe. Retrieved from https://www.webpagefx.com/blog/social-media/social-media-mobile-travel/.
50. Adukaite, A., Reimann, A. M., Marchiori, E., & Cantoni, L. (2013). Hotel mobile apps. The case of 4 and 5 star hotels in European German-speaking countries. In *Information and communication technologies in tourism 2014* (pp. 45–57). Cham: Springer.
51. Mistilis, N., & Gretzel, U. (2013). Tourism operators' digital uptake benchmark survey 2013. https://www.tra.gov.au/Archive-TRA-Oldsite/ Research/View-all-publications/All-Publications/tourism-operators-digital-uptake-benchmark-survey-2013-research-report

52. Kim, D.-Y., Park, J., & Morrison, A. M. (2008). A model of traveller acceptance of mobile technology. *International Journal of Tourism Research, 10*, 393–407.
53. Kwon, J. M., Bae, J., & Blum, S. C. (2013). Mobile applications in the hospitality industry. *Journal of Hospitality and Tourism Technology, 4*(1), 81–92.
54. Feinstein, E. (2016). How Mobile Payments Revolutionized the Travel and Hospitality Industry. Retrieved from https://mobilemarketingwatch.com/how-mobile-payments-revolutionized-the-travel-and-hospitality-industry-68103/.
55. Sorrells, M. (2018). What travel marketers need to know for digital engagement. Retrieved from https://www.phocuswire.com/Bing-Phocuswright-digital-travel-marketing-study.
56. Hutchinson, A. (2018). 5 Digital Trends to Watch in Hospitality Marketing [Infographic]. Retrieved from https://www.socialmediatoday.com/news/5-digital-trends-to-watch-in-hospitality-marketing-infographic/520225/.
57. Lee, K., Lee, H. R., & Ham, S. (2013). The effects of presence induced by smartphone applications on tourism: Application to cultural heritage attractions. In *Information and communication technologies in tourism 2014* (pp. 59–72). Cham: Springer.
58. Koo, C., Shin, S., Kim, K., Kim, C., & Chung, N. (2013). Association for Information Systems AIS Electronic Library (AISeL) smart tourism of the Korea: A case study. Pacific Asia Conference on Information Systems (PACIS). Retrieved from https://pdfs.semanticscholar.org/8074/446789ac a29337218ffa0b6e822b3280ab00.pdf
59. Dickinson, J. E., Ghali, K., Cherrett, T., Speed, C., Davies, N., & Norgate, S. (2014). Tourism and the smartphone app: Capabilities, emerging practice, and scope in the travel domain. *Current Issues in Tourism, 17*(1), 84–101.
60. Oliveira, T., Thomas, M., Baptista, G., & Campos, F. (2016). Mobile payment: Understanding the determinants of customer adoption and intention to recommend the technology. *Computers in Human Behavior, 61*, 404–414.
61. Peng, R., Xiong, L., & Yang, Z. (2012). Exploring tourist adoption of tourism mobile payment: An empirical analysis. *Journal of Theoretical and Applied Electronic Commerce Research, 7*(1), 21–33.
62. Kounavis, C. D., Kasimati, A. E., & Zamani, E. D. (2012). Enhancing the tourism experience through mobile augmented reality: Challenges and prospects. *International Journal of Engineering Business Management, 4*, 1–6.
63. Fritz, F., Susperregui, A., & Linaza, L. T. (2005). Enhancing cultural tourism experiences with augmented reality technologies. *The 6th International Symposium on Virtual Reality, Archaeology and Cultural Heritage VAST.*
64. Goh, D. H., Lee, C. S., Ang, R. P., & Lee, C. K. (2010). Determining services for the mobile tourist. *The Journal of Computer Information Systems, 51*(1), 31–40.
65. Han, D.-I., Jung, T., & Gibson, A. (2013). Dublin AR: Implementing augmented reality in tourism. In *Information and Communication Technologies in Tourism 2014* (pp. 511–523). Springer.
66. Mitola, J., & Maguire, G. Q. (1999). Cognitive radio: Making software radios more personal. *IEEE Personal Communications, 6*(4), 13–18.
67. Thomas, R. W., Friend, D. H., DaSilva, L. A., & MacKenzie, A. B. (2006). Cognitive networks: Adaptation and learning to achieve end-to-end performance objectives. *IEEE Communications Magazine, 44*(12), 51–57.
68. Förster, A. & Murphy, A. L. (2007). FROMS: Feedback routing for optimizing multiple sinks in WSN with reinforcement learning. In *Proc. 3rd Int. Conf. Intelligent Sensors, Sensor Netw. Inf. Process. (ISSNIP).*
69. Di, M. & Joo, E. (2007). A survey of machine learning in wireless sensor networks. In *Proc. 6th Int. Conf. Inf., Commun. Signal Process.*
70. Perera, C., Zaslavsky, A., Christen, P., & Georgakopoulos, D. (2013). Context aware computing for the Internet of Things: A survey. *IEEE Communication Surveys and Tutorials, 99*, 1–41.
71. Simón, F. J. G., Narangajavana, Y., & Marques, D. P. (2004). Carrying capacity in the tourism industry: A case study of Hengistbury Head. *Tourism Management, 25*(2), 275–283.

72. Dergi, G. M. (2018). Antalya'daki belgeli tesis sayısı 805'e, yatak sayısı 590 bin'e ulaştı. https://www.gmdergi.com/arastirma-haberleri/antalyadaki-belgeli-tesis-sayisi-805e-yatak-sayisi-590-bine-ulasti/. Accessed 27 Jan 2019.
73. Çelik, P., & Topsakal, Y. (2017). Akıllı Turizm Destinasyonları: Antalya Destinasyonunun Akıllı Turizm Uygulamalarının İncelenmesi. *Seyahat ve Otel İşletmeciliği Dergisi, 14*(3), 149–166.
74. Soteriades, M. (2012). Tourism destination marketing: approaches improving effectiveness and efficiency. *Journal of Hospitality and Tourism Technology, 3*(2), 107–120.

A Hybrid Approach for Image Segmentation in the IoT Era

Tallha Akram, Syed Rameez Naqvi, Sajjad Ali Haider, and Nadia Nawaz Qadri

1 Introduction

The Internet of Things (IoT) has a potential to transform our world by intelligently connecting every device that touches both of our professional and everyday lives through a range of sensors, cloud computing, etc. and also has a capacity to unlock the door to the next industrial revolution. In the last decade, IoT has covered numerous domains including communication [1, 2], energy efficiency [3, 4], vehicular systems [5], smart farming [6], healthcare [7], computer vision [8], and to name a few. Detecting object of interest from an IoT-based camera device can be useful, especially under the scenario of identifying one dedicated class from a bulk of processed information.

Spectral clustering is a class of graph theoretic procedure, which is popular for finding natural groupings. Over the last decade, it has become a widely adopted tool for image segmentation problems via the normalized cut (NCut) criterion [9]. Spectral clustering popularity increased substantially among the researchers in the field of computer vision, bioinformatics, robotics, etc. The idea has proven itself to be an active area of research among machine learning community. The core reasons behind are its capability to accurately group complex structured data and its simplicity in implementation. Image segmentation using spectral clustering is not an expedient way of dealing large size images due to its high computational demand and memory requirements. To overcome such predicament, a dimensional reduction procedure is integrated as an effective preprocessor before calculating an affinity matrix – which combats the curse of dimensionality. In the situation where computational bottleneck is comprehended, a competent selection of suitable

T. Akram (✉) · S. R. Naqvi · S. A. Haider · N. N. Qadri
Department of Electrical and Computer Engineering, COMSATS University, Islamabad, Pakistan
e-mail: tallha@ciitwah.edu.pk

F. Al-Turjman (ed.), *Artificial Intelligence in IoT*,
Transactions on Computational Science and Computational Intelligence,
https://doi.org/10.1007/978-3-030-04110-6_5

Fig. 1 A pictorial illustration of weighted binary tree-based fast spectral clustering

preprocessor reduces the bottleneck risk by reducing the data structure size. Selection of preprocessor is a key task and has foremost effects on clustering.

In this work, we propose an unsupervised image segmentation technique using spectral clustering aimed at salient object extraction. Few challenges faced by image segmentation based on spectral clustering are its infeasibility to process large images due to high computational demand, memory requirements, and its sensitivity to irrelevant and noisy data. These challenges are successfully addressed in the proposed work of weighted binary tree-based fast spectral clustering (WBTFSC); sample results are demonstrated in Fig. 1. The algorithm integrates dimensionality reduction method into spectral clustering by introducing an effective preprocessor, which is comprised of two fundamental steps: (1) color quantization based on weighted binary tree and (2) unique pixels selection. Noisy images are also tested with proposed algorithm and showed exceptional results comparable to the other implemented techniques.

Many clustering techniques are strictly bound to Euclidean geometry, making assumption explicitly or implicitly to form convex regions in Euclidean space, compared to be more flexible and confined to wider geometry range of spectral clustering [10].

Color quantization is deemed to be a prerequisite of many color image segmentation algorithms, and we exploited this key to be an effective preprocessor [11–13]. The contribution is twofold: Firstly, we leveraged the fact that natural images include fewer consistent groups than pixels [14] and introduced an effective preprocessor of weighted binary partition tree for quantization. Secondly, selection of unique pixels from normalized image further reduces the data structure size before subjecting to spectral clustering. Labeling is finalized using fuzzy C-means soft clustering to exhibit fuzzy behavior and making it feasible for salient object extraction with fused clusters.

In spectral clustering, image is represented by a weighted graph, where each pixel is a node connected to other nodes with edges. Despite the fact, spectral clustering achieves accurate clustering and makes precise segments, it is still not considered to be a strong competitor to other clustering techniques [15, 16]. Spectral algorithm forms affinity matrix of size ($n \times n$), where n is the total number of pixels in the image, and computes eigenvectors of this affinity matrix with computational complexity of $O(n^3)$ and memory requirement of $O(n^2)$.

To improve the performance of spectral clustering, multiple schemes were proposed and implemented [17, 18]. One technique is the providence of original data representatives, which should have the capacity of retaining original values. Exploiting the subsampling of randomly selected data w.r.t. to grouping criteria is another reasonable approach. Low-rank approximation in spectral clustering is another central option for the dimensionality reduction. High-quality clustering achieved by spectral clustering can be annihilated with poor choice of preprocessor, which must be selected carefully for improved performance.

2 Related Work

Spectral clustering for image segmentation is a graph theory-based information extraction procedure which describes the image as a weighted graph and partitions them using optimized cost function. Segmentation is done under the potency of feature vectors (directly/indirectly), based on eigendecomposition of graph Laplacian matrix [19]. Possibly the first comprehensive effort was made by Shi and Malik [20] to segment image using spectral clustering. They formulated the problem from a graph theoretic perspective and introduced a normalized cut to segment the image.

Traditionally spectral clustering performance degraded hastily when dealing with high-resolution images due to its high computational cost. To overcome this problem, few techniques are implemented depending on the nature of the problems. Bruce et al. [18] used the multilevel approach for graph partitioning in which sequence of increasingly smaller graphs approximate a complete graph. The smallest graph is then partitioned using spectral clustering and finally back propagated to graph hierarchy. Same spirit of multilevel approach is discussed by Yan et al. [21] where coarsening and un-coarsening methodology are followed and algorithm is designed to deal with the computational requirements to overcome spatial bottleneck. They employed fast approximate spectral clustering using K-means clustering and random projection trees as a preprocessing step. The intension was to minimize data reduction effect on clustering accuracy, in the spirit of rate-distortion theory.

Another approach was put forth by Ducoumau et al. [22], where clustering was configured as hypergraph cut problem to optimize the objective function. The proposed method was based on the principle of multilevel paradigm with three key steps, starting with the initial step of hypergraph reduction, followed by a spectral clustering of reduced hypergraph, and the final step of cluster refinement. Similarly, Tung et al. [23] implemented a new distinctive approach of block-wise processing in combination with stochastic ensemble consensus. They first divided the image into multiple nonoverlapping fix-sized blocks and then calculated affinity matrix for each block. Finally, they merged the segments using stochastic ensemble consensus.

Tasdemir et al. [26] proposed approximate spectral clustering in combination with neural networks by employing self-organizing maps and neural gas as quantizer – a preprocessing step. They exploited local density-based similarity measure without user-predefined parameters. Similarly, a fast affinity propagation clustering approach was implemented by Shang et al. [20] in which they considered both local and global structural information. The concept was based on multilevel graph partitioning which managed to implement both vector-based and graph-based clustering.

Fowlkes et al. [14] made an exertion to overcome the time and computational limitations by incorporating the low-rank approximation. They introduced Nystrom-based spectral clustering algorithm for image segmentation. The main idea was to solve the grouping problems for small random subsets of pixels in image and then anticipate this solution to all pixels. To improve Nystrom spectral clustering, few methods were also proposed in literature by applying different quantization schemes [27]. All techniques referred above are based on the concept of quantization before object of interest detection. A concept of finding salient regions is also presented which is based on attention mechanism and considers contrast to be a primary factor.

Goferman et al. [28] proposed context-aware saliency detection. The objective is to detect the salient region(s) with respect to a scene. They considered both low level and global feature. Cheng et al. [29] proposed histogram-based contrast method to measure saliency. One of the biggest challenges in saliency detection is to acquire salient regions w.r.t. scene understanding.

3 Clustering Methods

The proposed design is a combination of multiple clustering methods including hierarchical clustering (weighted binary partitioned tree), graph clustering (spectral), and finally fuzzy clustering (C-means). Sample results of proposed technique are demonstrated in Fig. 2.

3.1 Weighted Binary Partition Tree

Let us consider an image S of size $(n \times n)$ for $(m = n)$ where $s = (r, c)$ for $(s \in S)$ to be row and column indices. Initialize a binary tree clustering by partitioning S into

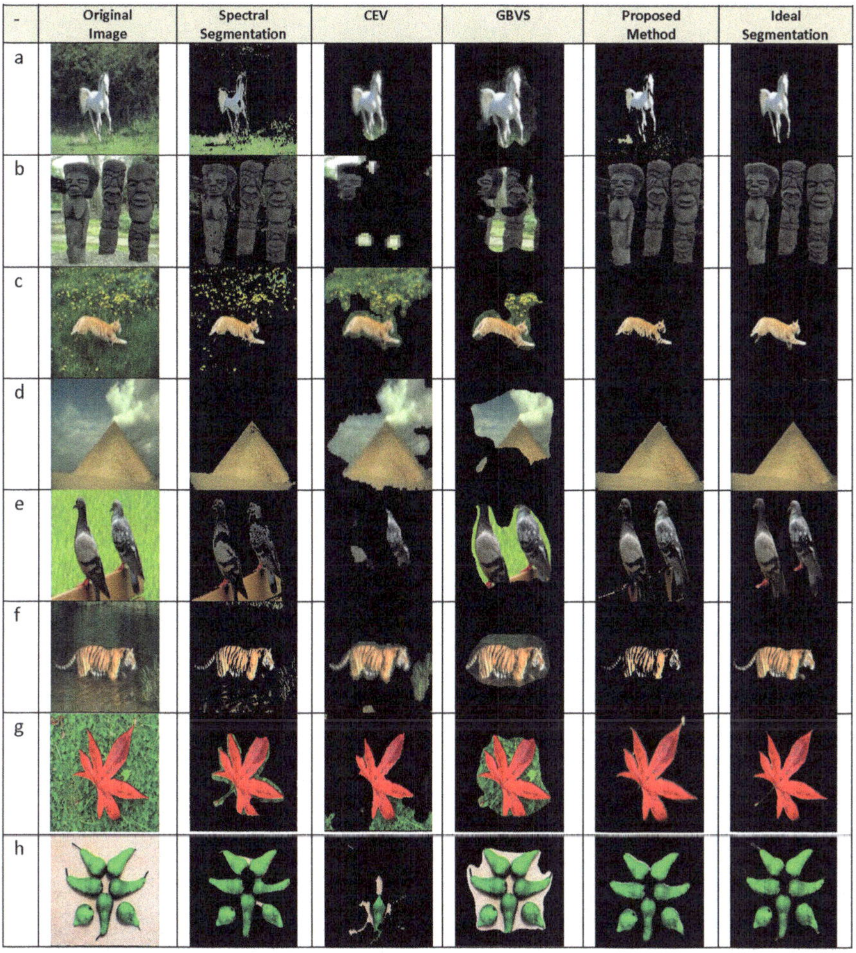

Fig. 2 Visual comparisons with respect to object of interest detection; (1) original image, (2) spectral algorithm for object extraction, [1] (3) curve evolution algorithm using low depth field, [17] (4) graph-based visual saliency [18], (5) WBTFSC algorithm for object extraction, (6) ground truth

K^1 disjoint sets, subject to the constraint of achieving maximum of two nodes (*left and right*), where the numbers of clusters are already defined.

Each partitioned node belongs to its parent node which flows up in the hierarchy to the root node "X" – the starting point of any partition. Let P_n represents set of image pixels which correspond to node n. For the proposed weighted binary tree, let the head node (n) be represented by a number "1," while the children nodes are $2n$ and $2n + 1$. The partition of X is on the basis of colors for the selected image. The partition is controlled by cluster colors which rely on mean and variance value and stored as linked list. In splitting process node, order is very important; following the procedure in an order, we need to state the second-order statistical properties.

$$R_n = \sum_{s \in P_n} W_s x_s x_s^t \tag{1}$$

$$O_n = \sum_{s \in P_n} W_s x_s \tag{2}$$

where minimum squared deviation is given by a cluster mean; therefore, it favors an assumption that mean is equal to q – the cluster's quantization value.

$$\mu_n = O_n / \mid P_n \mid \tag{3}$$

Cluster covariance is defined as:

$$\sum_{n+1} = \sum_{n} - \frac{1}{\mid P_n \mid} O_n O_n^t \tag{4}$$

Weights are assigned according to relation:

$$W_s = \left(\frac{1}{k_\sigma {}^* \min \left(\left\| \nabla S_{xy} \right\|, 14 + 2 \right)} \right)^2 \tag{5}$$

where K_σ is a Gaussian smoothing kernel which filters gradient estimator and S_{xy} is color range with minimum values of 2, the lowest visually indistinguishable level.

This approach is modified to show improved performance for large clusters, which endorses the maximum cluster's variation by handling and splitting order. Unit vector $\widehat{u}$ is defined as:

$$\sum_{s \in P_n} \left((x_s - \mu_s)^t \widehat{u} \right)^2 = \widehat{u}^t \sum_{n+1} \widehat{u} \tag{6}$$

The eigenvector $\widehat{u}$ corresponds to the largest eigenvector λ_n of $\sum_{n+1}$. Total squared variation in direction of $\widehat{u}$ is given as:

$$\sum_{s \in P_n} \left((x_s - \mu_s)^t \widehat{u} \right)^2 = \lambda_n \tag{7}$$

The points in the Wp_n are divided into Wp_{2n} sorted as Wp_{2n+1}

$$Wp_{2n} = \left\{ s \in P_n : \widehat{u}_n{}^t x_s \leq \widehat{u}_n{}^t \mu_n. \right.$$

and Wp_{2n+1} is defined as:

$$Wp_{2n+1} = \left\{ s \in P_n : \widehat{u}_n{}^t x_s > \widehat{u}_n{}^t \mu_n. \right. \tag{8}$$

New weighted statistics are created as:

$$R_{2n} = w_n R_n - w_{2n} R_{2n}$$

$$O_{2n+1} = w_n O_n - w_{2\text{n}} O_{2n} \tag{9}$$

$$P_{2n+1} = |w_n P_n| - |w_{2n} P_{2n}|$$

where the weights w_n and ω_{2n} are static to be 0.6 and 0.4, respectively, for optimal performance. The optimal splitting order is selected with the objective of maximum reduction in error squared, defined as:

$$\text{TSE} = \sum_{\text{All Leaves } n} \sum_{s \in p_n} \|x_s - \mu_n\| \tag{10}$$

Here we utilized the largest principle eigenvector to achieve the best split (Fig. 3).

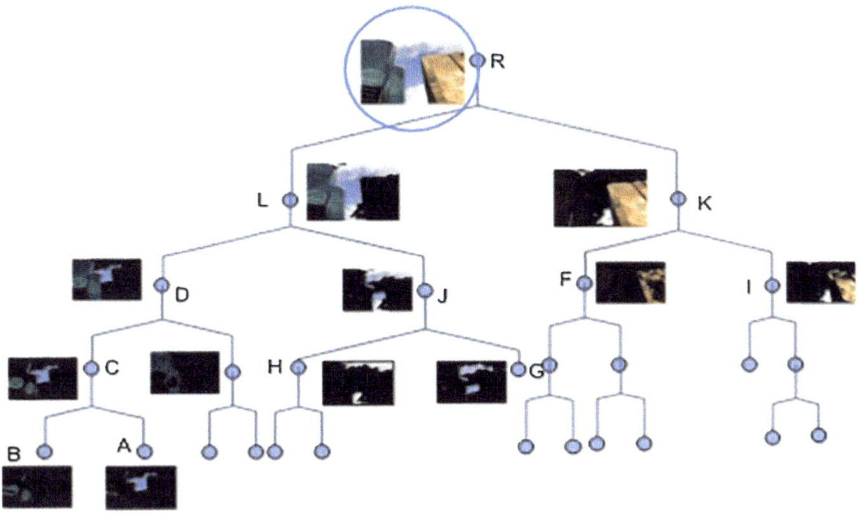

Fig. 3 Cluster partitioning using weighted binary partition tree

3.2 Spectral Clustering

Spectral clustering for image segmentation is a graph theory-based information extraction procedure which describes the image as weighted graphs and partitions them using optimized cost function. Let$G(V, E)$ be an undirected graph, where $V = (v_1, v_2, v_3, \ldots, v_n)$ are a set of n nonempty vertices along with set of edges, E. The graph is weighted; each edge between vertices v_i and v_j have a weight of $w_{ij} > 0$. G^t is subgraph of G having vertices and edges (V', E') with the fact $V' \in V$ & $E' \in E$. The weight adjacency is given as $W = (w_{ij})_{n \times n}$, where $w_{ij} \in \Re^{(n \times n)}$. For weight values $(w_{ij}) = 0$, correspond to the fact that vertices V_i and V_j are not connected.

Different techniques like $K-$neighbors, $k-$nearest neighbors, and fully connected graph are used to transform given data points $(x_1, \ldots, x_n)$ into pairwise distance d_{ij} or pairwise similarities s_{ij} into graph. Let d_{ij} be the dissimilarity matrix; the weight matrix $W = (w_{ij})$ for $(i,j) = 1, \ldots, n$ can be calculated from sigmoid function as:

$$w_{ij} = \exp - \left\| x_i - x_j \right\|^2 / 2\sigma^2 \text{ For } i \neq j \ \& \ 0 \text{ Otherwise.} \tag{11}$$

The parameter σ has supplementary impact on clustering obtained. The selection of σ is a very critical step; one approach is to run an algorithm for different values of σ until the least squared intra-cluster distance to its center position is obtained [23]. Mentioned hit and trial rule for optimal value of σ looks not feasible for large datasets, so few other strategies are instigated [31]. We selected sigma value 1 for the optimized solution.

The degree matrix D is defined as diagonal matrix with $D = [d_{ij}]_{n \times n}$, where $d_{ij} \in \Re^{n \times n}$ defined as $d_i = \sum_{j=1}^{n} w_{ij}$. un-normalized graph Laplacian is given as $L = D - W$, which may vary in case of unweighted graph as $L = D - A$, where A is adjacency binary matrix of G given as $A = a_{ij}(n \times n)$, where $a_{ij} = 1$ if edge connection is found between V_i and V_j and 0 if otherwise. The Laplacian matrix must satisfy the following properties. For vector $X \in \Re^{n \times n}$:

$$x^T L x = \frac{1}{2} \sum_{i,j=1}^{n} w_{ij} \left(x_i - x_j \right)^2 \tag{12}$$

L is symmetric positive definite. The smallest eigenvalue of L is 0, with corresponding eigen indicator vector 1, i.e., $1 = (1, 1, \ldots, 1)^T$. Luxburg [32] theoretically proves all the properties of graph Laplacian matrix. Fiedler suggested that λ_2 is the second eigenvector of L that represents relationship among connected graph components also known as fielder's eigenvector. The normalized Laplacian matrix is given as $L = I - D^{-(0.5)} W D^{-(0.5)}$. Second random-walk Laplacian matrix is given as $L_{rw} = I - D^{-1} W$.

4 Proposed Methodology

The proposed object extraction algorithm is summarized with flowchart shown in Fig. 4, which incorporates three fundamental stages; the first stage deals with color quantization using weighted binary tree, where the first loop divides the image into distinct clusters, while the second loop

assigns each cluster a unique color value from the palette. The second stage deals with the normalization procedure and finding unique pixels in the quantized image and followed by final stage of object of interest detection and later extraction using spectral clustering.

Algorithm analysis revealed us that reduction of a color palette size with weighted binary partition tree quantization still is not very effective in calculating reduced similarity matrix of size ($n \times n$), considerably when image is of large size. The reduction of distinct colors as an outcome of binary tree quantization has no effect on total number of nodes subjected to spectral clustering but to the number of natural colors. To overcome such predicament, all pixels are normalized in the range of ($0 \rightarrow 1$) before selecting unique values to construct similarity graph. Essentially, duplication removal process keeps in record total number of duplicate pixels and their respective addresses for each basic RGB value. This procedure allows us to deal with large size images in calculating affinity matrix, which is one of the prime steps in spectral clustering.

The number of clusters K is defined by user under constraint that the weighted binary tree quantization clusters are greater than spectral clustering, so not to annihilate the visual contents of image but to reduce the number of distinct colors. If number of clusters created using spectral clustering is insufficient in number, it will result in fused segments of object and background, resulting in false segmentation.

For similarity matrix, sigmoid function approach is used instead of K-nearest neighbors and ϵ-neighborhood graph, due to large number of connected components. Ultimately, grouping is done using fuzzy C-means algorithm, and cluster members are mapped to their respective addresses with the duplication sequence. The method shows performance degradation if preprocessor is not defined well or image having uniformly distributed pixels.

Based on the evaluation of WBTFSC in obtaining meaningful segmentation for object of interest detection, spectral clustering fulfills two conditions:

- Nodes that represent possibly larger connected regions belong to object of interest class and should be selected.
- Regions represented by selected clusters should be disjoint.

Fig. 4 General flow diagram of weighted binary tree-based fast spectral clustering

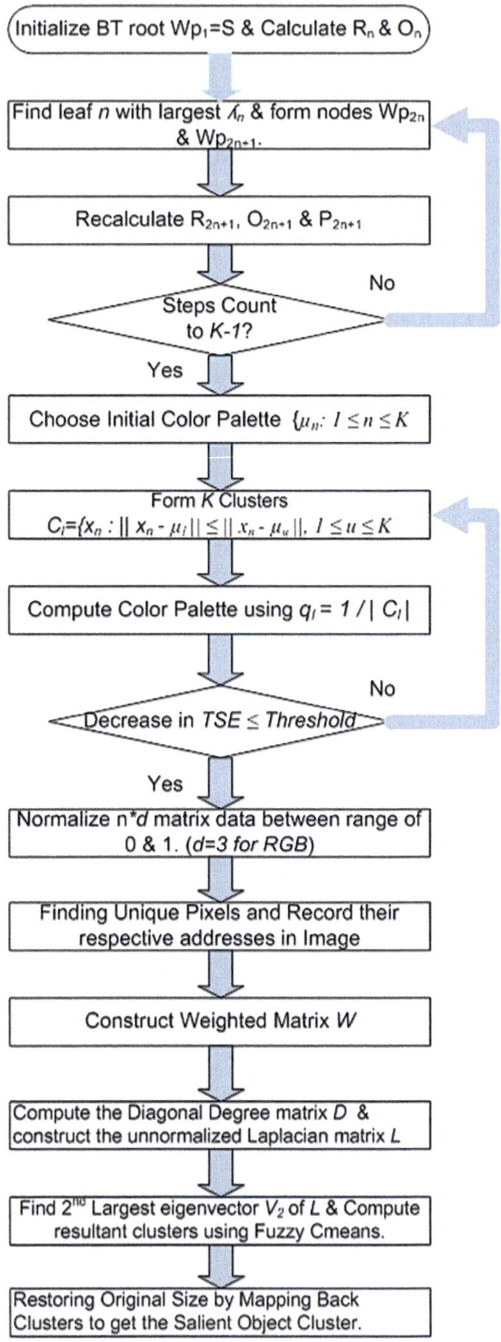

5 Results and Discussion

To assess the performance of proposed algorithm from the perspective of object of interest detection and extraction, a set of images are selected from MSRA[1] salient object database, Berkeley[2] segmentation dataset, and image collection of our group, with the perspective of finding at least one obvious salient object. To show the consistency and effectiveness of our proposed algorithm, performance measure was credited in respect of accuracy and efficiency. Results are visually compared with three state-of-the-art algorithms. Analysis is made from the perspective of object of interest extraction. In order to show the consistency and effectiveness of the proposed WBTFSC algorithm, different performance measures are credited, in respect to time, size, and efficiency. To demonstrate the performance of our proposed algorithm, we compared our results with three state-of-the-art algorithms, spectral clustering [30], curve evolution with low depth of field [24], and graph-based visual saliency (GBVS) [25].

Few images are selected for visual comparison between proposed technique and selected methods as shown in Figs. 2, 5, and 6. Ideal segmentation or ground truth is shown in column 6 which are manually segmented images to calculate accuracy. Our approach has shown notable results, visually observed in column 5, when compared with conventional spectral clustering [31] in column 2, curve evolution approach in column 3, and graph-based visual saliency in column 4. It is observed that not only the test images with single salient object showed good performance but test images with multiple salient objects are also segmented and extracted successfully and can be observed in images "b," "e," "I," "n," etc.

With weighted binary tree palette design, few image pixels are misclassified and covered with wrong cluster labels which results in the loss of object information with spectral clustering object extraction procedure. In order to coup with this situation, we took advantage of soft clustering of fuzzy C-means algorithm. It assigns data member x_t not to strictly bind to one cluster, rather deal it in flexible way. For the proposed technique, object extraction principle is based on color distinctness where each color represents different cluster. The presence of multiple clusters within salient object results in the loss of inner details of object and can be observed in images "f," "k," and "p," but objects' contours remain preserved.

Table 1 shows time comparison of all selected techniques and size comparison of proposed method and spectral clustering, which can be analyzed with respect to similarity graph time and total processing time to extract an object.

It can be noticed that WBTFSC has shown a remarkable reduction in similarity graph time to create $(n \times n)$ matrix. The maximum time for spectral method is 45.27 s for image "e" as compared to 9.28 s for WBTFSC. Minimum time for

[1] www.research.microsoft.com/en-us/um/people/jiansun/SalientObject/salient_object.htm

[2] www.elib.cs.berkeley.edu/segmentation

Fig. 5 Few more visual comparisons with respect to salient object extraction. (1) Original image; (2) spectral algorithm for object extraction [1]; (3) curve evolution algorithm using low depth field [17]; (4) graph-based visual saliency [18]; (5) WBTFSC algorithm for object extraction; (6) ground truth

WBTFSC is 1.30 s for "p" image compared to spectral clustering, 19.17 s and curve evolution approach, 34.31 s.

GBVS has shown minimum time for object detection, but results are in contrary to our approach in terms of accuracy. A comparison of proposed method and spectral clustering in terms of executaion time and similarity graph size is given in Fig. 7. Notable difference can be observed in execution time and similarity graph size where WBTFSC maximum nodes are approximately 16 K in size and execution time less than 20 s in contrast to spectral clustering.

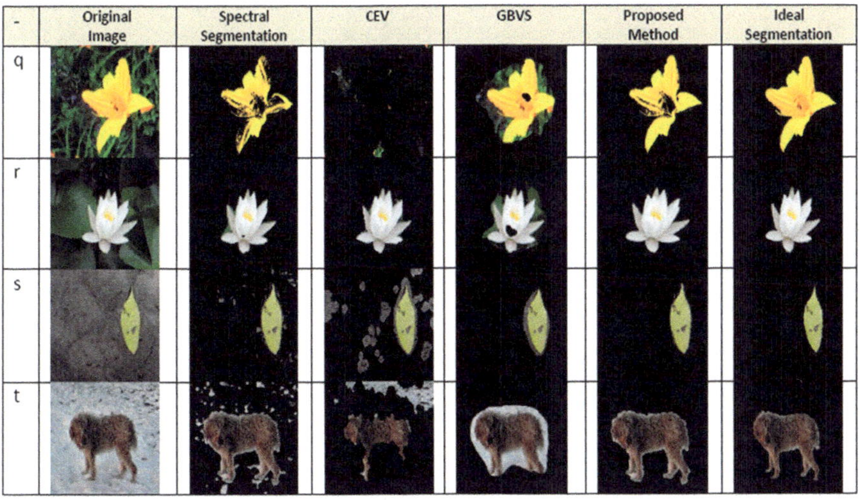

Fig. 6 Few more visual comparisons with respect to salient object extraction. (1) Original image; (2) spectral algorithm for object extraction; (3) curve evolution algorithm using low depth field; (4) graph-based visual saliency; (5) WBTFSC algorithm for object extraction; (6) ground truth

Spectral clustering performance degraded hastily when showed up with noise. This deficiency is challenged with our proposed method and showed exceptional results, demonstrated in Fig. 8. We introduced three different types of noises with different intensity levels and extracted the object of interest with good accuracy. It can be observed that our proposed technique showed better results especially when salt-and-pepper noise is introduced. Table 2 shows similarity graph time and total time for noisy images. For each column assigned to different noises, left sub-column shows our proposed method processing time, and right sub-column shows spectral clustering's processing time.

In order to objectively evaluate the segmentation performance from the perspective of object of interest detection and extraction, we consider two different strategies. For the first, output image is segmented into regions (R_1,R_2,...,R_n). $R_i \cap R_j = \phi$ for $i \neq j$ and $\cup_{i=1}^{n} R_i = I$. For each segment R_i in image, if more than 50% overlap with ground truth foreground O_{GT}, it counts as foreground R [34–35].

We tested the efficiency of images by dividing into 164 regions/segments; comparative plot is shown in Fig. 9 (right).

$$R_{\text{obj}} = \cup_{(i:(pR_i, O_{MS})>0.5)} R_i. \tag{13}$$

or

$$p\Big(R_i, O_{MS} \geq p\left(R_k, O_{MS}\right), \forall k \; \varepsilon \; [1, n_s] \tag{14}$$

Table 1 Time and size parameters for the proposed technique of WBTFSC in comparison with curve evolution method, GBVS, and spectral clustering

–	Spectral method			CEV	CBVS	BTFSC			
Figure#	S. graph	T. time	S. graph size	T. time	T. time	S. graph	T. time	Q-B-tree	S. graph size
a	41.50	42.85	18406	38.13	1.62	1.84	2.65	0.051	3680
b	29.62	30.60	15397	28.08	1.31	7.50	8.60	0.048	7234
c	33.13	34.85	16538	16.87	1.25	0.92	1.65	0.050	2630
d	26.93	27.61	13548	24.96	1.21	6.55	6.64	0.041	6973
e	43.16	45.27	19341	37.57	1.24	8.25	9.28	0.042	8393
f	30.05	31.64	15673	41.78	1.23	7.84	9.11	0.043	8310
g	33.01	37.40	17234	49.28	1.21	7.20	7.31	0.042	7589
h	39.03	42.42	18301	43.58	1.27	7.69	7.73	0.042	7654
i	26.93	27.75	13610	48.15	1.27	5.32	5.38	0.043	4701
j	14.10	15.19	9461	16.56	1.19	4.61	5.50	0.045	4768
k	22.65	23.19	12865	41.8	1.32	2.94	3.80	0.045	3584
l	18.22	21.26	12268	29.86	1.39	2.20	3.15	0.041	2951
m	6.67	7.07	7186	14.28	1.37	1.34	1.38	0.042	2601
n	31.63	33.10	16005	30.30	1.25	3.32	4.19	0.041	3470
o	22.42	23.09	12838	45.98	1.25	17.11	18.60	0.048	11345
p	18.58	19.17	11258	34.31	1.22	1.25	1.30	0.050	2544
q	15.73	16.10	10496	30.87	1.30	2.67	2.71	0.041	3590
r	17.05	21.49	12605	41.44	1.38	2.50	3.40	0.056	3162
s	42.02	44.26	19289	76.44	1.19	17.37	18.88	0.048	11470
t	19.73	20.38	11531	27.21	1.22	8.96	9.96	0.051	8471

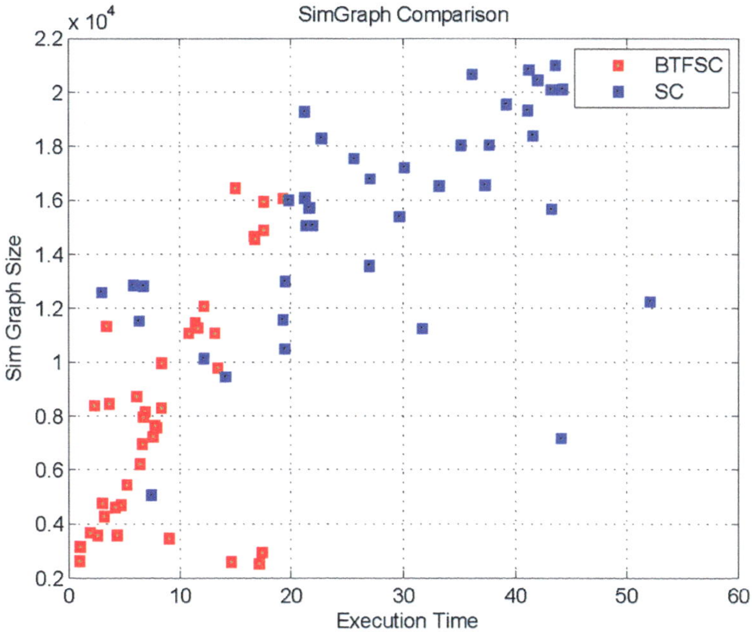

Fig. 7 Execution time vs. similarity graph size. A comparison between proposed method of WBTFSC and spectral clustering for the set of image samples

Fig. 8 Visual comparison of noisy images. Gaussian and salt-and-pepper noise are introduced on test images [column 1, column 4], and results can be visually analyzed using spectral algorithm and WBTFSC; (a) original image with Gaussian noise; (b) spectral algorithm; (c) proposed algorithm; (d) original image with salt-and-pepper noise; (e) spectral algorithm; (f) proposed algorithm

where

$$p\left(R_i, O_{GT}\right) = \max\left\{\frac{\mid R_i \cap O_{GT} \mid}{\mid R_i \mid}, \frac{\mid R_i \cap O_{GT} \mid}{\mid O_{GT} \mid}\right\} \quad (15)$$

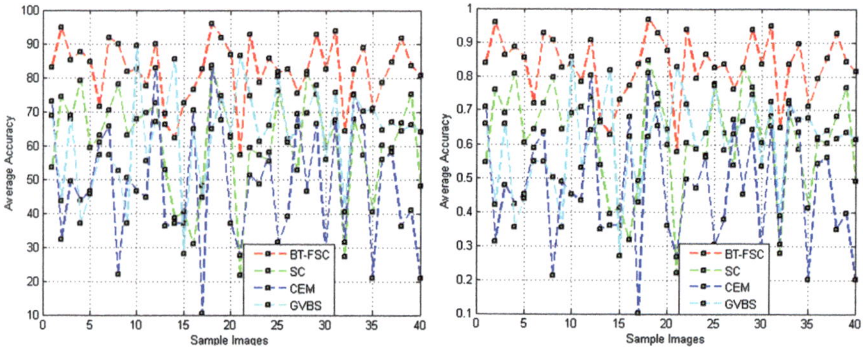

Fig. 9 Performance measure of proposed approach and three state-of-the-art methods with two different techniques. (Left) proposed approach for calculating accuracy, (right) using approach [25]

For the second approach, consider same test image I with background O' where $O' \varepsilon I$ and foreground salient object O for $(O \in I)$. The idea is to compare background of ground truth O'_{GT}, with the background of proposed algorithm O'_{PA}, and foreground of ground truth O_{GT} with foreground of proposed algorithm O_{PA}. The reason to compare background is to determine the background fragments eradication percentage.

Any change of pixel value greater than selected threshold is considered in calculating efficiency. Percentage is calculated by counting the color variations to the total number of background pixels O'_{GT} and O'_{GT}. Foreground efficiency of O_{GT} and O_{PA} is calculated by considering all the parameters including the size, shape, and color of the object in ground truth and proposed algorithm. The efficiency results are shown in Fig. 9 (left).

Multiple techniques are adopted to ensure performance measure, e.g., precision and recall (Fig. 10) and mean absolute error (MAE) computed with optimum threshold based on ground truth mask.

Precision and recall is calculated as:

$$\text{Precision} = \frac{\sum_{i=1}^{W} \sum_{j=1}^{H} \quad S_{TH}(i, j) \quad S_{GT}(i, j)}{\sum_{i=1}^{W} \sum_{j=1}^{H} S_{TH}(i, j)} \tag{16}$$

Recall is calculated as:

$$\text{Recall} = \frac{\sum_{i=1}^{W} \sum_{j=1}^{H} \quad S_{TH}(i, j) \quad S_{GT}(i, j)}{\sum_{i=1}^{W} \sum_{j=1}^{H} S_{GT}(i, j)} \tag{17}$$

Table 2 Time parameters for test images with noise. In each partition, "left" column shows WBTFSC similarity graph times and total time, whereas "right" column shows spectral similarity graph time and total time

–	Gaussian noise				Salt-and-pepper noise				Speckle noise			
Fig#	SG.T	T.T	SG.T	T.T	SG.T	T.T	SG.T	T.T	SG.T	T.T	SG.T	T.T
1	3.27	4.26	52.63	56.56	2.87	3.80	44.68	46.90	2.44	3.81	51.06	53.40
2	18.77	20.59	53.49	57.90	19.18	21.27	41.97	44.21	25.24	27.66	52.29	55.60
3	12.45	14.24	54.64	60.08	8.79	10.[illegible]64	31.96	33.32	9.28	11.06	49.73	54.13
4	0.40	1.29	55.48	59.95	0.34	1.2[illegible]	36.28	38.53	0.31	1.33	49.75	54.23

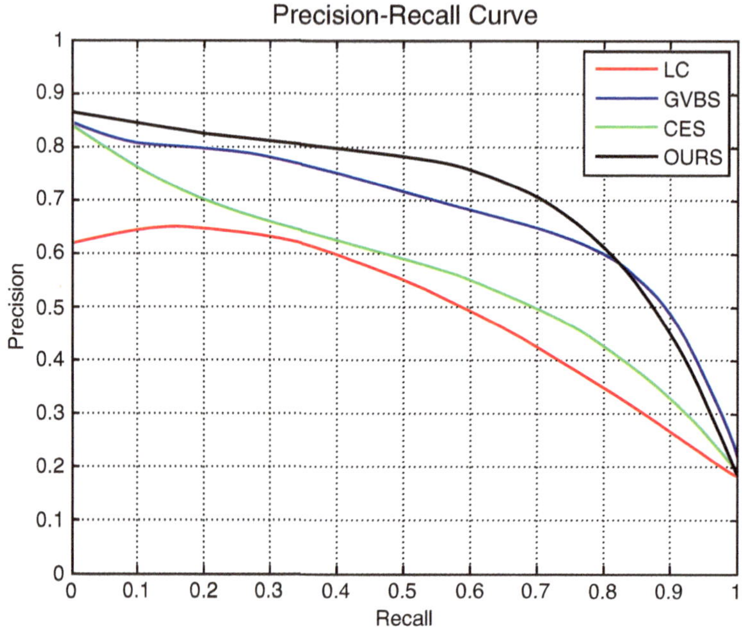

Fig. 10 Precision and recall curve of proposed method on selected datasets

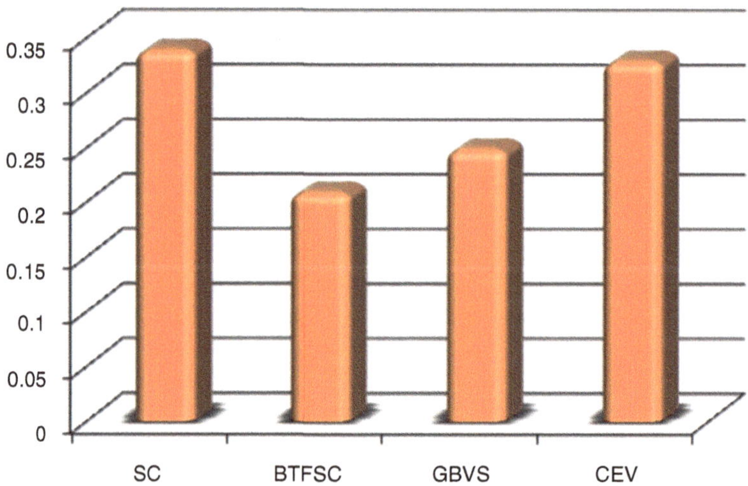

Fig. 11 Mean absolute error of various extraction methods w.r.t. ground truth

where S_{TH} threshold generated binary object mask corresponds to saliency map and S_{GT} is its corresponding ground truth (Fig. 11).

For our experiments to calculate precision, we selected fixed thresholding, respectively, for generating binary object mask.

MAE [36] is calculated as:

$$\mathrm{MAE} = \frac{1}{W \times H} \sum_{i=1}^{W} \sum_{j=1}^{H} \mid S_{\mathrm{Extract}}(i, j) - S_{Gt}(i, j) \mid \tag{18}$$

W and H are the parameters showing image width and height, respectively. Lower MAE value indicates better performance, which provides a better estimate of dissimilarity between the proposed extraction method and ground truth.

For the proposed method, simulation is done using MATLAB (Ver. 8.0) using Pentium Core 2 Quad processor, 3GHz.

6 Chapter Summary

In this chapter, we presented a cascaded design for object of interest detection and finally extraction using weighted binary tree quantization and spectral graph method. The first level of hierarchy reduced distinct number of pixels available in natural scenes using weighted binary tree quantization. The next stage identifies unique color pixels in the image which further reduce number of pixels, as spectral clustering constructs affinity matrix of size ($n \times n$), where n is the total number of pixels in image.

Selection of unique pixels reduces number of nodes for graph method to overcome memory limitations and computational time. The results are subjected to fuzzy C-means for final grouping, followed by a concluding step of mapping back pixels to their original addresses. The algorithm showed satisfactory results compared to other state-of-the art algorithms. Noisy images are also tested, Gaussian, salt-and-pepper, and speckle, and achieved amazing results compared to other object extraction approaches. In addition to design's effectiveness, there are few shortcomings which need to be addressed in future articles including number of predefined clusters for both weighted binary tree and spectral clustering. Additionally, few conditions should be considered in order to identify the objects of interest as it shows different color and contrast compared to the background, different textural details, etc.

References

1. Al-Turjman, F. (2018). QoS–aware data delivery framework for safety-inspired multimedia in integrated vehicular-IoT. *Elsevier Computer Communications Journal, 121*, 33–43.
2. Al-Turjman, F., & Alturjman, S. (2018). 5G/IoT-enabled UAVs for multimedia delivery in industry-oriented applications. *Multimedia Tools and Applications, Springer*. https://doi.org/10.1007/s11042-018-6288-7.
3. Alabady, S., & Al-Turjman, F. (2018). A novel approach for error detection and correction for efficient energy in wireless networks. *Springer Multimedia Tools and Applications*. https://doi.org/10.1007/s11042-018-6282-0.

4. Hasan, M. Z., & Al-Turjman, F. (2018). Analysis of cross-layer design of quality-of-service forward geographic wireless sensor network routing strategies in green internet of things. *IEEE Access Journal, 6*(1), 20371–20389.
5. Al-Turjman, F., & Alturjman, S. (2018). Confidential smart-sensing framework in the IoT era. *The Springer Journal of Supercomputing*. https://doi.org/10.1007/s11227-018-2524-1.
6. TongKe, F. (2013). Smart agriculture based on cloud computing and IOT. *Journal of Convergence Information Technology., 8*(2), 210–216.
7. Bhatt, C., Dey, N., & Ashour, A. S. (Eds.). (2017). *Internet of things and big data technologies for next generation healthcare* (pp. 978–973). Cham: Springer.
8. Han Zou, Yuxun Zhou, Jianfei Yang, Costas J. Spanos, "Device-free occupancy detection and crowd counting in smart buildings with WiFi-enabled IoT", Energy and Buildings, V(174), 309–322, 2018.
9. Ng, A. Y., Jordan, M. I., & Weiss, Y. (2002). On spectral clustering: Analysis and algorithm [C]. NIPS.
10. Hartmann, S. L., & Galloway, R. L. (2000). Depth-buffer targeting for spatially accurate 3-D visualization [J], medical images. *IEEE Transactions on Medical Imaging, 19*(10), 1024–1031.
11. Lei, T., & Sewchand, W. (2001). Object detection and recognition via stochastic model-based image segmentation [C]. IEEE Multidimensional Signal Processing Workshop, 2001, Sixth, pp. 17–18.
12. Yixin Chen, J. (2002). A region based fuzzy feature matching a roach to content-based image retrieval [J]. *IEEE Transactions on Fuzzy Systems, 24*(9), 1252–1267.
13. Liu, Z., Shen, L., & Zhang, Z. (2011). Unsupervised image segmentation based on analysis of binary partition tree for salient object extraction [J]. *Signal Processing, 91*(2), 290–299.
14. Fowlkes, C., Belongie, S., Chung, F., & Malik, J. (2004). Spectral grouping using the Nystrom method [J]. *IEEE Transaction on Pattern Analysis and Machine Intelligence, 26*(2), 214–225.
15. Itti, L., Koch, C., & Niebur, E. (1998). A model of saliency-based visual attention for rapid scene analysis [J]. *IEEE Transactions on Pattern Analysis and Machine Intelligence, 20*(11), 1254–1259.
16. Mohanta, P. P., Mukherjee, D. P., & ST, A. (2002). Agglomerative clustering for image segmentation [C]. *IEEE International Conference on Pattern Recognition, 1*, 664–667.
17. Zhao, P., & Zhang, C. Q. (2011). A new clustering method and its a lication in social networks [J]. *Pattern Recognition Letters, 32*(15), 2109–2118.
18. Hendrickson, B., & Leland, R.(1995). A multilevel algorithm for partitioning graphs [C]. *IEEE Conference on SuperComputing*.
19. Fiedler, M. (1973). Algebraic connectivity of graphs. *Czechoslovak Mathematical Journal, 79*, 57–70.
20. Shi, J., & Malik, J. (2000). Normalized cuts and image segmentation [J]. *IEEE Transactions on Pattern Analysis and Machine Intelligence, 22*(8), 888–905.
21. Yan, D., Huang, L., & Jordan, M. (2009). Fast a roximate spectral clustering [R]: Technical report UCB/EECS]-2009-45.
22. Ducournau, A., Bretto, A., Rital, S., & Laget, B. (2012). A reductive a roach to hypergraph clustering: An a lication to image segmentation [J]. *Pattern Recognition, 45*(7), 2788–2803.
23. Tung, F., Wong, A., & Clausi, D. A. (2010). Enabling scalable spectral clustering for image segmentation [J]. *Pattern Recognition, 23*, 4069–4076.
24. Gao, H., Mei, J., & Si, Y. (2013). A curve evolution a roach for unsupervised segmentation of images with low depth of field [J]. *IEEE Transaction in Image Processing, 22*(10), 4086–4095.
25. Harel, J., Koch, C., & Perona, P. (2006). Graph-based visual saliency [J]. *Advances in Neural Information Processing Systems, 19*, 545–552.
26. Tasdemir, K. (2010). Vector quantization based on proximate spectral clustering of large datasets [J]. *Pattern Recognition, 45*(8), 3034–3044.
27. Zhang, K., Tsang I. W., & Kwok, J. T. (2008) Improved Nystrom low-rank a roximation and error analysis [C]. *International Conference on Machine Learning. ICML* (pp. 1232–1239).
28. Goferman, S., Zelnik-Manor, L., & Tal, A. (2010). Context-aware saliency detection [C]. In *Proc. IEEE Int. Conf. Comput. Vis. Pattern Recognit.*, 2010, pp. 2376–2383.

29. Cheng, M. M., Zhang, G. X., Mitra, N. J., & Huang, X. (2011). Global contrast based salient region detection [C]. In *IEEE Conference on Computer Vision and Pattern Recognition*, June 20, 2011.
30. Ng, A. Y., Jordan, M. I., & Weiss, Y. (2001) On spectral clustering: Ananlysis and an algorithm [J]. *Advance in Neural Information Processing Systems,* 849–856.
31. L.Z. Manor, & P. Perona. (2004). Self tuning spectral clustering [J]. *Advance in Neural Information Processing Systems* (pp. 1601–1608).
32. Luxburg, U. (2007). A tutorial on spectral clustering [J]. *Journal of Statistics and Computing, 17*(4), 395–416.
33. Ge, F., Wang, S., & Liu, T. (2007). New benchmark for image segmentation evaluation [J]. *Journal of Electronic Imaging, 16*(3), 033011.
34. Clinton, N., Holt, A., & Gong, P. (2010). Accuracy assessment measures for object-based image segmentation goodness [J]. *Photogrammetric Engineering & Remote Sensing, 76*(3), 289–299.
35. Duan, Q., Akram, T., Duan, P., & Wang, X. (2016). Visual saliency detection using information contents weighting. *Optik-International Journal for Light and Electron Optics, Elsevier, 127*(19), 7418–7430.
36. Khan, Z. U., Akram, T., Naqvi, S. R., Haider, S. A., Kamran, M., & Muhammad, N. (2018). Automatic detection of plant diseases; utilizing an unsupervised cascaded design. (IBCAST), Islamabad, pp. 339–346, 3028.

Big Data Analytics for Intelligent Internet of Things

Mohiuddin Ahmed, Salimur Choudhury, and Fadi Al-Turjman

1 Introduction

A successful of Internet of Things (IoT) environment requires standardization that contains interoperability, compatibility, reliability, and effectiveness of the operations on a global scale [1]. The rapid growth of cloud computing facility and the IoT causes a sharp growth of data. Enormous amounts of networking sensors are continuously collecting and transmitting data to be stored and processed in the cloud. Such data can be environmental data, geographical data, astronomical data, logistic data, etc. Mobile devices, transportation facilities, public facilities, and home appliances are the primary data acquisition equipment in IoT. The volume of such data will surpass the capacities of the IT architectures and infrastructure of existing enterprises and, due to real-time analysis character, will also greatly impact the computing capacity [2]. Management of these increasingly growing data is a challenge for the community in general. Figure 1 shows a year over year rise on the amount of data.

Due to the characteristics of the data being generated from IoT environment, we can call these data as Big data. The challenges faced by the IoT users compelled to label these data as Big data! Therefore the Big data generated by IoT has

M. Ahmed
College of Technology and Design, Canberra Institute of Technology, Canberra, ACT, Australia
e-mail: m.ahmed.au@ieee.org

S. Choudhury
Department of Computer Science, Lakehead University, Thunder Bay, ON, Canada
e-mail: salimur.choudhury@lakeheadu.ca

F. Al-Turjman (✉)
Department of Computer Engineering, Antalya Bilim University, Antalya, Turkey
e-mail: fadi.alturjman@antalya.edu.tr

F. Al-Turjman (ed.), *Artificial Intelligence in IoT*,
Transactions on Computational Science and Computational Intelligence,
https://doi.org/10.1007/978-3-030-04110-6_6

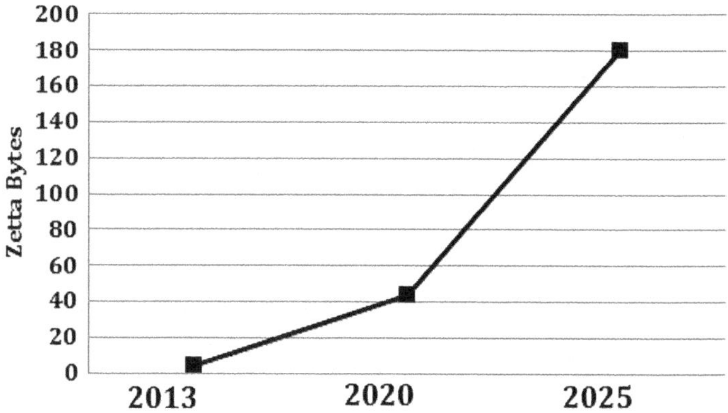

Fig. 1 Year over year rise of data volume. 1 Zetta Byte = 1 Trillion Gigabytes [3]

unique characteristics due to different types of data collected. The most common characteristics of these data reflect heterogeneity, variety, unstructured feature, noise, and high redundancy [2, 4]. It is envisaged that by 2030, the quantity of sensors will reach one trillion, and then the IoT data will be the most important part of Big data, according to the forecast of HP [5]. In fact, various artificial intelligence (AI) techniques have been applied to wireless sensors to improve their performance and achieve specific goals. We can look at AI techniques as a means of introducing an intelligent learning in a key enabling technology in IoT, which is the wireless sensor network (WSN). Learning is an important element in the observe, analyse, decide, and act (OADA) cognition loop [6, 7], used to implement the idea of cognitive wireless networks [8, 9]. We can broadly classify AI techniques as computational intelligence (CI) techniques, reinforcement learning (RL) techniques, cognitive sensor networks and multi-agent systems (MAS), and context-aware computing. Although these techniques are closely related with each other, we can segregate them to show the different goals that learning can achieve for the network.

CI techniques are a set of nature-inspired computation methodologies that help in solving complex problems that are usually difficult to fully formulate using simple mathematical models. Examples of CI techniques include genetic algorithms, neural networks, fuzzy logic, simulated annealing, artificial immune systems, swarm intelligence, and evolutionary computation. In a learning environment, CI techniques are useful when the learning agent cannot accurately sense the state of its environment. In WSNs, CI techniques have been applied to problems such as node deployment planning, task scheduling, data aggregation, energy-aware routing, and QoS management. Authors in [10] have provided an extensive survey of CI techniques applied to WSNs. They elaborate on various CI techniques and associate each with typical problem domains they can solve in WSNs. From their observations, swarm intelligence applied to solving the routing and clustering problem has drawn the most research attention in recent times. However, a major drawback of this methodology is that it can be computationally intense and may require some form of model-based offline learning to deliver to the requirements of the application scenario. Techniques such as ant colony optimization can cause an undesirable increase in communication overhead in WSNs [11] too. Apart from these drawbacks, none of the CI algorithms have been applied to solving problems of data representation, aggregation, and delivery in a distributed, decentralized setup, under dynamic communication constraints, as is the case in data hungry IoT applications.

According to Intel [1, 12], data produced from IoT has three distinguishing features:

- Terminals generating massive amount of data;
- Semi-structured or nonstructured;
- Data of IoT is not useful without analysis.

Due to the generation of Big data by IoT, the existing data processing capacity of IoT is becoming ineffective, and it is imperative to incorporate Big data technologies to promote the development of IoT. It is important to understand that the success of IoT lies upon the effective incorporation of Big data analytics. The widespread deployment of IoT also gives a challenge to Big data community to propose newer techniques as Big data and IoT are interdependent. On one hand, the widespread deployment of IoT produces data both in quantity and category, thus providing the opportunity for the application and development of Big data; on the other hand, the incorporation of Big data analytics to IoT simultaneously accelerates the research advances and business models of IoT.

Figure 2 shows a holistic view of the relationship between IoT and Big data. Earlier the amount of data generated by IoT systems could be easily handled by the traditional data analytics techniques. However, as amount of data being generated by the IoT systems are transformed into Big data, the traditional analytics methods

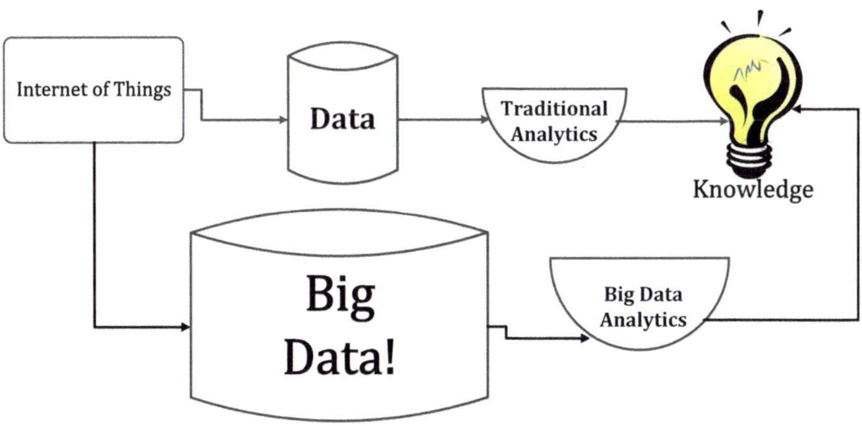

Fig. 2 Association relation between IoT and Big data

are not effective. It is important to understand the fact that, the traditional data analytics methods are not suitable to extract knowledge from Big data. For gaining meaningful insights or knowledge from the Big data generated by IoT, a set of robust data analytics method is required. Therefore, it is imperative to incorporate the analytics techniques suitable for Big data generated by the IoT systems.

In this chapter, the key aspects of Big data analytics are presented in detail. Section 2 starts with the definition of Big data; Sect. 3 reflects the challenges associated with Big data. Section 4 provides the taxonomy of Big data analytics followed by subsections on data acquisition & storage, programming model, benchmarking, analysis, and applications. Section 5 contains the future research directions of IoT data analytics followed by conclusion of the chapter.

2 Definition of Big Data

Big data is an abstract concept [13]. The concept of Big data is rudimentarily dependent on the configuration of a system, i.e. RAM, HDD capacity, etc. [14]. The significance of Big data has been recognized very recently and has different opinions on its definition. In layman's term, Big data reflects the datasets that could not be perceived, acquired, managed, and processed by traditional IT and software/hardware tools in efficient manner [15]. Communities like scientific and technological enterprises, research scholars, data analysts, and technical practitioners have different definitions of Big data [16]. The following set of definitions provides an understanding on the profound social, economic, and technological connotations of Big data:

- Apache Hadoop [17]: "Datasets which could not be captured, managed, and processed by general computers within an acceptable scope".
- McKinsey & Company [18]: "Big data shall mean such datasets which could not be acquired, stored, and managed by classic database software".
- Doug Laney [19]: "Challenges and opportunities brought about by increased data with a 3Vs model, i.e., the increase of Volume, Velocity, and Variety".
- IBM, Microsoft [20, 21]: "In the 3V model, Volume means, with the generation and collection of masses of data, data scale becomes increasingly Big; Velocity means the timeliness of Big data; Variety indicates the various types of data, which include semi-structured and unstructured data such as audio, video, webpage, and text, as well as traditional structured data".
- NIST [22]: "Big data means the data of which the data volume, acquisition speed, or data representation limits the capacity of using traditional relational methods to conduct effective analysis or the data which may be effectively processed with important horizontal zoom technologies, which focuses on the technological aspect of Big data".
- Manyika et al. [23]: "Datasets whose size is beyond the ability of typical database software tools to capture, store, manage, and analyze".

- Davis and Patterson [24]: “Big data is data too Big to be handled and analyzed by traditional database protocols such as SQL”.
- Edd Dumbill [25]: “Explicitly conveys the multi-dimensionality of Big Data when adding that the data is too Big, moves too fast, or doesn’t fit the strictures of your database architectures”.

To describe Big data, a number of V have been used in the literature. Here we combine all the Vs as below:

- Volume: The term “volume” is related with the amount of data and its dimensionality. The advantage from the ability to process large amounts of information is the main attraction of Big data analytics. The consequence is that it is a trend for many companies to store vast amount of various sorts of data: social networks data, healthcare data, financial data, biochemistry and genetic data, astronomical data, etc.
- Variety: “Variety” refers to the mix of different types of data. These data do not have a fixed structure and rarely present themselves in a perfectly ordered form and ready for processing [26]. Indeed, such data can be highly structured, semi-structured, or unstructured (video, still images, audio, clicks, etc.).
- Variability: It can be added to “variety” to emphasize on semantics, or the variability of meaning in language and communication protocols.
- Velocity: “Velocity” involves streams of data, structured records creation, and availability for access and delivery. Most of the Internet-based applications are streaming in nature and a source of Big data. The importance is reflected in the speed of the feedback loop, taking data from input through to decision.
- Value: This feature is the purpose of Big data technology. This view is well expressed by the International Data Corporation [4] when saying that Big data architectures are “designed to economically extract value from very large volumes of a wide variety of data, by enabling high-velocity capture, discovery, and/or analysis”. This value falls into two categories: analytical use and enabling new business models, products, and services [17–25].
- Veracity: Veracity defines the truth or fact, or in short, accuracy, certainty, and precision. Uncertainty can be caused by inconsistencies, model approximations, ambiguities, deception, fraud, duplication, incompleteness, spam, and latency. Due to veracity, results derived from Big data cannot be proven, but they can be assigned a probability [4].

3 Challenges with Big Data

The Big data era brings a set of challenges on data acquisition, storage, management, analysis, and so on. Traditional data management and analysis systems are based on the relational database management system (RDBMS) [4]. However, such systems are usable with structured data only and cannot deal with the semi-structured or unstructured data which are a significant portion of Big data. In fact, Big data are not necessarily structured data and require preprocessing before analysis. It is visible that the traditional RDBMSs cannot handle the huge volume and heterogeneity of Big data. The research community has proposed some solutions from different perspectives. A promising solution is cloud computing which is utilized to meet the requirements on infrastructure for Big data, i.e. cost-efficiency and smooth upgrading/downgrading, etc. In a nutshell, it is a non-trivial task to deploy Big data analysis infrastructure. The key challenges are listed as follows.

3.1 Data Representation

The collected data from different sources are composed of certain levels of heterogeneity in the type, structure, semantics, organization, granularity, and accessibility. Therefore, it is important to properly represent Big data for further analysis. The goal of proper data representation is to make data more meaningful for analysis and user interpretation. An improper data representation will significantly impact the value of the original data and barrier to effective data analysis. An example of efficient data representation contains data structure, class, and type, as well as integrated technologies to enable efficient operations on different datasets [4, 26].

3.2 Redundancy

Usually the collected data comes with a high level of redundancy. For effective data analysis, it is important to use redundancy reduction and data compression approaches. For example, a large portion of data generated by sensor networks is highly redundant, which are required to be filtered and compressed for a robust analysis. In Big data environment, newer technologies are required to be incorporated as the redundant Big data will have a significant impact on the analysis [4, 26].

3.3 Privacy and Security

Most Big data service providers or owners outsource their datasets for effective maintenance and analysis due to their limited capacity. Therefore, usage of external bodies or tools increases the potential privacy and safety risks. For example, the transactional dataset contains details of the lowest granularity and sensitive information such as credit card numbers. Therefore, outsourced analysis of Big data is only recommended with proper preventive measures such as data anonymization of sensitive data, to ensure its security and privacy [4, 26].

3.4 Energy Efficiency

The energy consumption of high-end computing facilities is alarming due to their impact on both economic and environmental perspectives. Needless to mention that in the Big data environment, the energy consumption will be much higher than before and is unexpected from both financial and environmental perspectives. The technology industry is looking for green computing; however, the Big data is a main constraint for this venture. Therefore, it is urgent to devise new approaches to control power consumption and management mechanism without affecting the expandability and accessibility are ensured [4, 26].

3.5 Challenges with Big IoT Data

Smart cities are constructed in IoT paradigm. Therefore, Big data originates from a number of sectors such as industry, agriculture, traffic, transportation, healthcare, public departments, and so on [1–4]. According to the data acquisition and transmission approach in IoT, its network architecture may be divided into three layers: the sensing layer, the network layer, and the application layer. The sensing layer is responsible for data acquisition and mainly consists of sensor networks. The network layer is responsible for information transmission and processing, where close transmission may rely on sensor networks, and remote transmission shall depend on the Internet. Finally, the application layer support specific applications of IoT. The challenges associated with the Big IoT data are summarized as below:

- Large-scale data: Plenty of data acquisition sensors are distributed which acquire heterogeneous data.
- Strong time and space correlation: The time and space correlation is an important property of IoT data. During data analysis and processing, time and space are also important dimensions for statistical analysis.
- Effective data: Unexpected and huge amount of noises usually occur during the acquisition and transmission of data in IoT. Among datasets acquired by acquisition devices, only a small amount of abnormal data is valuable.

4 Taxonomy of Big Data Analytics

Figure 3 shows the taxonomy of Big data analytics. It consists of five basic aspects of Big data. The first category, Big data acquisition and storage, covers data acquisition and storage management. The Big data programming model includes research on programming models used in the Big data environment. The benchmark process covers the evaluation of Big data systems. The Big data analysis involves studies which focus on approaches to extract knowledge from Big data. The final category,

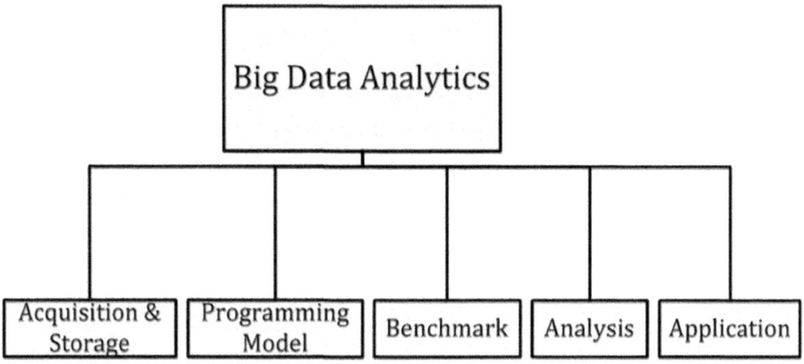

Fig. 3 Big data analytics taxonomy

application, covers studies related to the applications of Big data analytics in social, scientific, and business domains along with IoT which is the main focus of this chapter.

4.1 Acquisition and Storage

Big data comes along a new set of challenges to collect and store data efficiently. Compared to existing methods, Big data analytics need to deal with huge amounts of heterogeneous and unstructured data. It is not a trivial task to collect, integrate, and store these data by traditional data mining techniques and relevant infrastructures. These phenomena led the researchers to open a newer branch called Big data acquisition and storage. Next, the Big data acquisition and storage management is discussed.

4.1.1 Big Data Acquisition

Data acquisition is the process to aggregate information in a well-organized digital form for further storage and analysis. It is a combination of data collection, data transmission, and data preprocessing. Due to the fastest-growing sensor technology such as the Internet of Things (IoT) and radio-frequency identification (RFID), sensor-based data collection has become quite popular data acquisition approach [1–4]. However, due to the high initial investment for installation and maintenance creates a barrier in the Big data environment.

To avoid high expenditure, crowd-driven data collection was suggested by some researchers as an alternative to sensor-based data collection. The incorporation of the crowd workers in the data acquisition process helps reduction of noisy data and the collection of new types of data. There are a number of examples of crowd-driven data acquisition which are summarized below:

- FixMyStreet [27]: Users can specify the spatial location of any given street on a map and report problems associated with the location.
- Ushahidi [28]: Provides real-time data collection by enabling data collection from different channels such as email, social media, etc.
- EcoTop [29]: Reduces the noise during data collection by issuing rewards to the mobile crowd workers.

These platforms provide incentives for peer collaboration among the crowd workers to achieve data availability and quality [4]. After the data collection step, Big data is required to be transmitted to the data centre for cleaning, processing, and integration. The transmission of Big data is posed to a set of challenges such as input/output bottlenecks, network traffic delays, and data replication [4]. To overcome these challenges, researchers adopt various approaches to improve the efficiency of Big data transmission [13]. Another major challenge of Big data analytics is the integration of heterogeneous unstructured data collected from different sources. Data accessibility, common data platform, and consolidated data model were identified as three key levels of data integration [4]. Many researchers have proposed their approaches and platforms based on these levels.

4.1.2 Big Data Storage

The next step after Big data acquisition (combination of collection and transmission) is storage. The main functionality of data storage is to store and manage Big datasets with reliability and availability [4]. Infrastructure and data storage management are two basic parts of data storage [17].

- Infrastructure: Traditional infrastructure for data storage includes random access memory (RAM), magnetic disks, and storage class memory [20, 21]. Due to the specific performance of these infrastructures, it is a challenge to combine all these for Big data environment. The transmission of large amounts of data from hard disks to memory often limits the performance of Big data analytics [22]. Considering a network architecture, the data storage infrastructure can be categorized [18] as direct-attached storage (DAS), network-attached storage (NAS), and storage area network (SAN). These architectures are unable to support Big data analytics. However, storage virtualization proposed by Hasan and Al-Turjman [30] offers a way to accommodate the requirements of Big data analytics. Storage virtualization is the combination of multiple network storage devices that become a single storage unit [31] and allows Big data to be easily searched and linked through a single source. Thus, data can be transferred consistently regardless of the physical infrastructure reducing the cost of storage and easier Big data analysis [32].
- Data storage management: Data storage management focuses on the file systems and database technology [33]. Google designed the Google File System (GFS) for large distributed data-intensive applications [34]. By reducing the cost of hardware, it is able to provide fault tolerance and high performance [4]. GFS

lacks efficiency for small-sized files, and some other systems such as Hadoop Distributed File System [35], Kosmos distributed file system [36], and few others were developed to fulfil the requirements of Big data storage. The variety and volume features of Big data are the important challenges to traditional relational database systems. None Structured Query Language (NoSQL) is a new type of database modelled using means other than the tabular relations [37] where the key characteristics are partition tolerance, high availability [4]. Therefore, NoSQL is a good solution for Big data storage management. Most popular NoSQL databases are SimpleDB, Cassandra, HBase, Bigtable [38], and MongoDB [39]. Other than these solutions for storage management for Big data, many researchers proposed their own solutions based on these NoSQL databases.

4.2 Programming Model

Big data processing is the next challenge after handling the storage issue. According to Pino et al. [40], there are four primary requirements involved in Big data processing as shown in Fig. 4. A number of solutions are available to fulfil these requirements. Specifically, the available programming models are designed to map applications to the parallel environment. Traditional parallel models lack the scalability and fault tolerance required by Big data [26]. These led to the development of new architectures like MapReduce [40], PreGel, GraphLab, Dryad [41], and so on. Most popular model is the MapReduce paradigm due to its robust Big data handling approaches. We briefly discuss the MapReduce below.

MapReduce is a Big data programming model that uses a wide variety of clusters to achieve automatic parallel processing and distribution. The computing

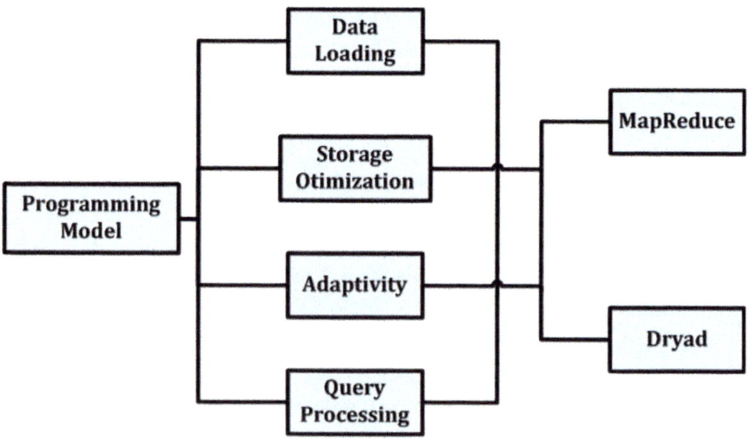

Fig. 4 Big data programming models

model contains a map function and a reduce function. The job of map function is to partition large computational tasks into smaller tasks and assign them to the appropriate key pair. After obtaining the output of the map function, the reduce function merges all values which share the same key value and generates a set of merged output values. The basic idea of MapReduce is to split a Big task into several chunks and execute the chunks in parallel to optimize. In the MapReduce model, the user only needs to focus on these above-mentioned two functions (map and reduce). As a popular and powerful programming model, MapReduce has been widely deployed to solve Big data-related problems such as in distributed computation, online aggregation, database system optimization, high-performance computing, DNA sequencing, text analytics, and many more.

Dryad is another framework for parallel applications [41] similar to MapReduce. In this model, a task is represented as a directed acyclic graph (DAG) which includes vertices and channels. Dryad completes the task by executing the vertices of the graph on a set of high-end computers and communicating through data channels. As a variant of MapReduce, Dryad generalizes MapReduce to arbitrary DAGs. This feature makes Dryad more flexible for the Big data applications with different structures [4].

One negative aspect of the MapReduce and Dryad models is that some agendas like behaviour abstraction, application optimization, and system simulation and migration are not well approached. These lead to a need for a generalized model that can bridge applications and various software frameworks for Big data analytics.

4.3 Benchmark

Advent of Big data in the scientific arena brings forth a newer set of benchmarking techniques among the researchers. The benchmarking techniques developed so far can be classified into two groups (component benchmarks and system benchmarks), and a simple taxonomy is shown in Fig. 5. The component benchmark has a limited scope and evaluates the performance of components in a Big data environment [4]. On the other hand, system benchmark focuses on the performance evaluation of an entire system [4].

Different types of benchmarking systems are discussed and summarized briefly below:

- PigMix [42], GridMix [43], GraySort [44]: The Standard Performance Evaluation Corporation (SPEC)'s central processing unit (CPU) benchmark [4]
- TeraSort [45] TeraSort (benchmark for Apache Hadoop system [17]) measures the amount of time to sort a large amount of distributed data in a given system. It has three parts as generation, sorting, and validation. The generation part creates random data. The sorting part performs the sorting and writes sorted data to Hadoop's distributed file system. The validation function reads sorted data to verify if it is in order.

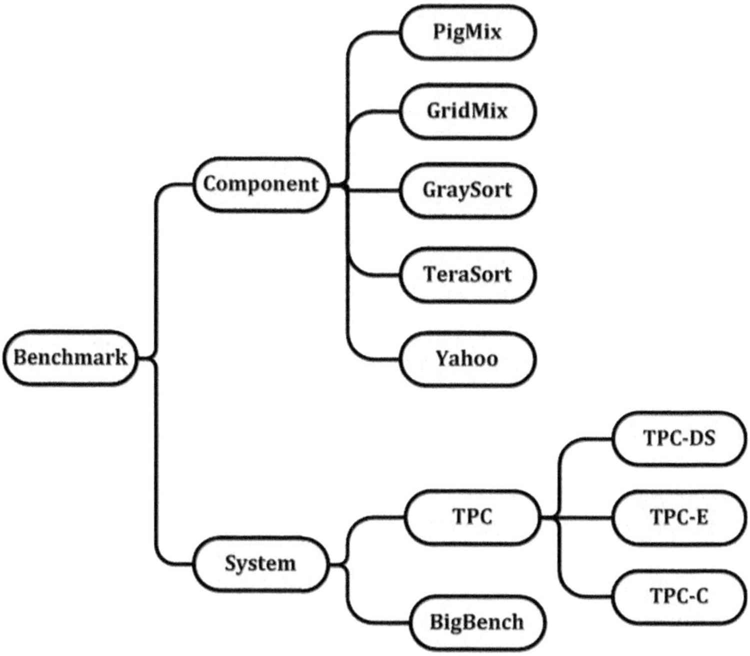

Fig. 5 Big data benchmarking

- Yahoo! [46] Yahoo! Cloud Serving Benchmark (YCSB) tries to bridge the gap of the existing benchmarking systems which were only designed for databases such as SQL. YCSB contains an extensible workload generator and several core workloads. The open-source YCSB workload generator can be used to load and execute various datasets and workloads. System Transaction Processing Performance Council [47] TPC-C, TPC-E, TPC-H, and TPC-DS are the perfect tools to measure a Big data system's ability for database loading and query executing.
- BigBench [48] It has two main components: (i) a data generator and (ii) query workload. The data generator helps to provide upgradeable volumes of structured, semi-structured, and unstructured data. The workload is devised with a set of queries that can cross different dimensions based on data.

4.4 *Big Data Analytics*

Data analytics is an important aspect of a successful IoT and in the dawn of Big data era; its importance has been being observed as never before. Organizations are interested in information from massive data to bring values. In this chapter, taxonomy is shown as in Fig. 6 based on the type of analysis being practised.

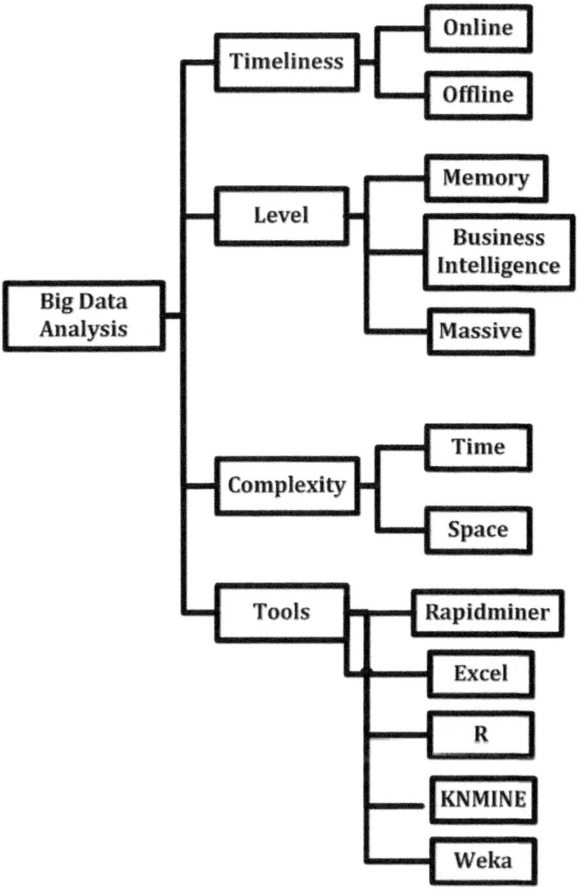

Fig. 6 Types of analytics

Because of the characteristics of Big data discussed in Sect. 3, different analytical architectures shall be considered for different application requirements.

4.4.1 Timeliness of Analysis

According to timeliness requirements, Big data analysis can be classified into online and offline analysis. These are discussed as follows:

– Online Analysis: From the IoT perspective, the online analysis is extremely important as the deployed sensors are constantly collecting data. Therefore from a particular time frame, it is necessary to analyse the data collected to reach out to any decision on anomalous event detection, cybersecurity assurance, etc. Also, in other applications such as e-commerce, financial trading data constantly changes, and rapid data analysis is needed. The widely accepted architectures for online analysis include (i) parallel processing clusters using traditional relational databases and (ii) memory-based computing platforms. For example, Greenplum from EMC and HANA from SAP are both real-time analysis architectures [4, 16].

- Offline Analysis: Offline analysis is usually required for applications without high requirements on response time, e.g. machine learning, statistical analysis, and recommendation algorithms [4, 16, 26]. Offline analysis carried out by retrieving data into a special platform through Big data acquisition tools. In the Big data environment, it is important to have specialized platforms to reduce the cost of data processing and improve the efficiency of data acquisition. Such platforms include the open-source tool Scribe from Facebook [4, 16], LinkedIn's open-source tool Kafka [4], Hadoop [17], and so on. These tools are capable of meeting the demands of offline data analysis with hundreds of MB per second.

4.4.2 Analysis at Different Levels

Big data analysis can also be classified into memory-level analysis, business intelligence (BI) level analysis, and massive level analysis, which are briefly discussed below:

- Memory-level analysis: When the data volume is smaller than the maximum memory of a cluster, this type of analysis is required. In recent times, the memory of server cluster surpasses hundreds of gigabytes. As a result, an internal database technology is advisable to use to improve the analytical efficiency. For online analysis, memory-level analysis is extremely suitable. A representative memory-level analytical architecture is MongoDB [39]. In the age of SSD (solid-state drive), the capacity and performance of memory-level data analysis have been further improved and widely applied.
- Business intelligence (BI) analysis: This analysis is required when the data scale surpasses the memory level; however, it can be imported into the BI analysis environment. The currently mainstream BI products are provided with data analysis plans to support the level over TB [4, 16].
- Massive analysis: When the BI analysis and traditional analysis are overwhelmed by the Big data, it is required to introduce the technologies like Hadoop and MapReduce to store and analyse the data. In recent times, most massive analysis utilizes HDFS of Hadoop to store data and use MapReduce for data analysis. Most massive analysis belongs to the offline analysis category.

4.4.3 Analysis with Different Complexity

The time and space complexity of Big data analysis algorithms varies from each other due to Big data characteristics (variety) and also application demands. For example, for the applications that require parallel processing, a distributed algorithm may be designed, and a parallel processing model may be used for data analysis [16, 26].

4.5 *Tools for Big Data Mining and Analysis*

A wide range of tools for Big data analytics are available which includes professional, expensive commercial software and also open-source software [4]. This section covers top five Big data analytics tool widely used in the community according to a survey of "What Analytics, Data mining, Big Data software that you used in the past 12 months for a real project?" of 798 professionals made by KDnuggets in 2012 [49].

4.5.1 R

R is a language and environment for statistical computing and graphics [50]. It is a predecessor of the S language and environment which was developed at Bell Laboratories. R can be considered as a different implementation of S. R provides a wide variety of statistical techniques including linear and nonlinear modelling, classical statistical tests, time-series analysis, classification, clustering, etc. and is highly extensible. Furthermore, in a survey of "Design languages you have used for data mining/analysis in the past year" in 2012, R was also in the first place, defeating SQL and Java. Due to the popularity of R, database manufacturers, such as Teradata and Oracle, have released products supporting R. Programming with Big data in R (pbdR) is a series of R packages and an environment for statistical computing with Big data by using high-performance statistical computation [4, 16]. The pbdR uses the same programming language as R with S3/S4 classes and methods which is used among statisticians and data miners for developing statistical software.

4.5.2 Excel

MS Excel is the most commonly used and powerful data processing and statistical analysis tool. Excel becomes a resourceful Big data analytics tool when some advanced plug-ins, such as Analysis ToolPak and Solver Add-in, with powerful functions for data analysis are integrated. This plug-ins can be used only if users enable them. Excel is also the only commercial software among the top five [49].

4.5.3 RapidMiner

RapidMiner [51] is an open-source software used for data mining, machine learning, and predictive analysis. However, to explore the Big data analytics, it is required to have licensed version which is not free. In an investigation of KDnuggets in 2011, it was more frequently used than R (ranked Top 1). Data mining and machine learning programs provided by RapidMiner include extract, transform, and load (ETL), data preprocessing and visualization, modelling, evaluation, and deployment. The data mining flow is described in XML and displayed through a graphic user interface (GUI). RapidMiner is written in Java. It integrates the learner and evaluation method of Weka and works with R. A handsome amount of regularly used data mining and machine learning algorithms can be implemented with connection of processes including various operators.

4.5.4 KNMINE

KNIME (Konstanz Information Miner) is a user-friendly and open-source data integration, processing, and analysis platform [52]. It provides the users to create data flows in a visualized manner and to selectively run analytical processes along with the analytical results, models, and interactive views. KNIME was written in Java and contains a large number of plug-ins. Through plug-in files, users can insert processing modules for files, pictures, and time series and integrate them into various open-source projects, e.g. R and Weka. In addition, it is easy to expand KNIME. Developers can effortlessly expand various nodes and views of KNIME.

4.5.5 Weka

Weka, developed by University of Waikato researchers, which comes from Waikato Environment for Knowledge Analysis, is an open-source machine learning and data mining software written in Java [53]. Weka provides almost all the fundamental functions such as data processing, feature selection, classification, regression, clustering, association rule, visualization, etc.

4.6 Applications of Big Data

The incorporation of Big data analytics makes it easier for organizations to gain meaningful information for being successful in their venture. Big data from various sources such as websites, emails, mobile devices, and social media are all important for knowledge extraction. In recent times, Big data has attracted a lot of attention from academia and industry which is easily understood by the amount of research paper published and job advertisements for data analyst/scientists. In [26], it is mentioned that the most common applications of Big data are in business, social, and scientific applications. A brief discussion on these areas along with IoT is given below.

4.6.1 Business Applications

There are huge amounts of operational and financial data stored at millions of different data sources in the business organizations. Big data analytics provides more agility to firms and makes it easier for firms to collect and analyse operational and financial data. Valuable information from Big data analytics allows managers to make more successful decision and identification of market conditions. Big data analytics have been already incorporated in various areas such as customer behaviour analysis, purchasing patterns, supply chain management, market forecasting, risk management, and fraud detection. In summary, Big data analytics empowers the business organizations to create new products and business processes, expand customer intelligence, and increase revenue.

4.6.2 Social Applications

Big data analytics facilitate information sharing in society, detect correlations among social events, and aggregate and analyse information to assist decision-making [26]. Out of many applications of Big data analytics for social good, a timely example of Big data analytics has been observed in election campaigns. With the help of Big data analytics, political data analytics advanced from simple micro targeting to true predictive data science, and the track record is good according to Jon Markman who is an investment adviser, trader, columnist, and author. In education sector, a number of research projects are ongoing to develop new algorithms for student behaviour analysis and effective mode of learning and teaching. Hidden patterns and trends identified using Big data analytics techniques can provide educators with valuable insights for the evaluation of the learning process. Additionally, human behaviours can be analysed through gathered Big data from a variety of sources. The modelling of human behaviours will allow governing body to interpret and predict social events such as traffic distribution, civil unrest incidents, and disease outbreaks. In the healthcare sector, Big data analytics provides the intelligence for electronic health records (EHRs) by connecting operational and clinical analytic systems and supports evidence-based healthcare. Evidence-based healthcare encompasses the systematic reviewing of previous clinical data in order to provide decision-makers with information as well as predictive analytics.

4.6.3 Scientific Applications

The continuous development of Big data analytics supports scientists to access large amounts of data quickly, facilitate data collection and sharing, and discover hidden patterns in Big datasets. Currently, Big data analytics has been applied in a lot of research areas. Especially in astronomy disciplines, Big data analytics has proven to be an efficient tool to address multiwavelength, multi-messenger, and huge amounts of astronomical data. NASA uses Big data analytics for real-time data processing on the flight operations. There are thousands of scientific applications where Big data analytics is being used and new inventions are in place.

4.6.4 Application of Intelligent IoT-Based Big Data

IoT is undoubtedly an important source of Big data and, simultaneously, one of the major market shares of Big data applications. As sensors are being used across almost every industry, the IoT is going to trigger a massive influx of Big data. IoT is going to have the biggest impact in the future of Big data analytics. A simple example may be given by logistic enterprises that may have profoundly experienced with the application of IoT Big data. In this scenario, the delivery vehicles of DHL may be equipped with sensors, wireless adapters, and GPS, so the headquarter is able to pinpoint the location. In addition to that application, supervision and management of employees can be executed with optimized performances.

5 Conclusion and Research Directions

In this chapter, the relationship between IoT and Big data is explored as the technological advances are inevitable. The Big data generated from IoT requires proper management and analysis to make IoT successful. Therefore, the importance of Big data analytics in IoT is a challenge. This chapter covers the Big data terminologies in the light of IoT and discusses the taxonomy of Big data analytics. Finally the chapter finishes with a set of research directions for the collaborative Big IoT data analytics. This chapter is going to be useful resource for anyone who is interested in IoT analytics and can be used as reference for graduate research students. In the next few subsections, the research direction of Big IoT data analytics is provided.

The Big data analytics research is in its early stage as the era just started few years ago and is confronting many challenges in different areas. Significant research efforts are required to improve the efficiency of Big data analytics. It is indeed an interesting research area with great potential, and there are many important problems to be solved by the collaboration of both academia and industry. There is a universal definition required for Big data. As observed in Sect. 2 of this chapter, the researchers/technology organizations are yet to reach to an agreed definition of Big data. An accepted formal definition is urgent for a variety of application domains to correlate Big data.

The presence of Big data challenges the traditional data management approaches. Currently, a plethora of research contributions on Big data technologies including data acquisition, storage, programming, benchmarking, and analytics are made. However, in the way the Big data is growing forth, it is imperative to continue research and development of relevant technologies.

As highlighted in the IoT perspective, the value gained from Big data is far higher than the value of non-Big dataset! As a result, the integration of different data sources (as in Big data) is a prerequisite for a successful venture nowadays. Moreover, the integration is posed too many challenges, such as different data patterns, redundant data, etc.

In Big data environment, the traditional security and privacy providers are insufficient. Since the data volume is fast growing, safety risks are more than ever before, and the existing security measures are not suitable for Big data. The Big data privacy is concerned with the protection of data acquisition patterns, i.e. personal interests, properties, etc. of users, and the privacy of data which may be leaked during storage, transmission, and usage, even if acquired with the permission of users. In a nutshell, as the Big data emerges, it is vital to ensure its security and privacy. A lack of data security and privacy can cause detrimental effects such as great financial losses and reputational damage for any organization.

References

1. Giusto, D., Iera, A., Morabito, G., & Atzori, L. (2010). *The internet of things: 20th Tyrrhenian workshop on digital communications*. New York: Springer Science & Business Media.
2. Li, S., Da Xu, L., & Zhao, S. (2015). The internet of things: A survey. *Information Systems Frontiers, 17*(2), 243–259.
3. Big Data: 20 Mind-Boggling Facts Everyone Must Read. (2015). https://www.forbes.com. [Online; accessed 29-August-2017].
4. Chen, M., Mao, S., & Liu, Y. (2014). Big data: A survey. *Mobile Networks and Applications, 19*(2), 171–209.
5. HP: Big Data Platform. (2017). http://www8.hp.com/us/en/software-solutions/Big-data-platform-haven/index.html. [Online; accessed 29-August-2017].
6. Haykin, S., et al. (2005). Cognitive radio: Brain-empowered wireless communications. *IEEE Journal on Selected Areas in Communications, 23*(2), 201–220.
7. Mitola, J., & Maguire, G. Q. (1999). Cognitive radio: Making software radios more personal. *IEEE Personal Communications, 6*(4), 13–18.
8. Zaki Hasan, M., & Al-Turjman, F. (2018). Swarm-based data delivery in social internet of things. In F. Al-Turjman (Ed.), *Smart things and femtocells* (pp. 179–218). Boca Raton: CRC Press.
9. Friend, D. H., Thomas, R. W., MacKenzie, A. B., & Silva, L. A. (2007). Distributed learning and reasoning in cognitive networks: Methods and design decisions. In Q. H. Mahmoud (Ed.), *Cognitive networks: Towards self-aware networks* (pp. 223–246). Hoboken: Wiley.
10. Al-Turjman, F. (2018). Fog-based caching in software-defined information-centric networks. *Computers & Electrical Engineering, 69*(1), 54–67.

11. Al-Turjman, F. (2017). Information-centric sensor networks for cognitive IoT: An overview. *Annals of Telecommunications, 72*(1), 3–18.
12. Alabady, S., & Al-Turjman, F. (2018). Low complexity parity check code for futuristic wireless networks applications. *IEEE Access, 6*(1), 18398–18407.
13. Liu, X., Iftikhar, N., & Xie, X. (2014). Survey of real-time processing systems for big data. In *Proceedings of the 18th international database engineering & applications symposium*, IDEAS'14 (pp. 356–361). New York: ACM.
14. Reed, D. A., & Dongarra, J. (2015). Exascale computing and big data. *Communications of the ACM, 58*(7), 56–68.
15. Fang, H., Zhang, Z., Wang, C. J., Daneshmand, M., Wang, C., & Wang, H. (2015). A survey of big data research. *IEEE Network, 29*(5), 6–9.
16. Chong, D., & Shi, H. (2015). Big data analytics: A literature review. *Journal of Management Analytics, 2*(3), 175–201.
17. Apache Hadoop. (2017). http://hadoop.apache.org/. [Online; Accessed 29-Aug-2017].
18. McKinsey & Company. (2017). http://www.mckinsey.com/. [Online; Accessed 29-Aug-2017].
19. Doug Laney. (2017). https://www.gartner.com/analyst/40872/Douglas-Laney. [Online; Accessed 29-Aug-2017].
20. What is Big Data. (2017). https://www.ibm.com/Big-data/us/en/. [Online; Accessed 29-Aug-2017].
21. Understanding Microsoft Big data solutions. (2017). https://msdn.microsoft.com/en-us/library/dn749804.aspx. [Online; Accessed 29-Aug-2017].
22. Big Data Information. (2017). https://www.nist.gov/el/cyber-physical-systems/Big-data-pwg. [Online; Accessed 29-Aug-2017].
23. Manyika, J., Chui, M., Brown, B., Bughin, J., Dobbs, R., Roxburgh, C., & Byers, A. H. (2011). Big data: The next frontier for innovation, competition, and productivity. Technical report, McKinsey Global Institute, June 2011.
24. Big ethics for Big data. (2017). https://www.oreilly.com/ideas/ethics-Big-data-business-decisions. [Online; Accessed 29-Aug-2017].
25. Planning for Big Data. (2017). http://www.oreilly.com/data/free/planning-for-Big-data.csp. [Online; Accessed 29-Aug-2017].
26. Ahmed, M., Anwar, A., Mahmood, A. N., Shah, Z., & Maher, M. J. (2015). An investigation of performance analysis of anomaly detection techniques for big data in scada systems. *EAI Endorsed Transactions on Industrial Networks and Intelligent Systems, 15*(3), 5.
27. FixMyStreet. (2017). https://www.fixmystreet.com/. [Online; Accessed 29-Aug-2017].
28. Ushahidi. (2017). https://www.ushahidi.com/. [Online; Accessed 29-Aug-2017].
29. Padhariya, N., Mondal, A., Goyal, V., Shankar, R., Madria, S. K. (2011). *EcoTop: An economic model for dynamic processing of top-k queries in mobile-P2P networks* (pp. 251–265). Berlin/Heidelberg: Springer.
30. Hasan, M. Z., & Al-Turjman, F. (2018). Analysis of cross-layer design of quality-of-service forward geographic wireless sensor network routing strategies in green internet of things. *IEEE Access, 6*(1), 20371–20389.
31. U.S. Patent No. 6,948,044. (2017). https://www.uspto.gov/. [Online; Accessed 29-Aug-2017].
32. Huber, N., Becker, S., Rathfelder, C., Schweflinghaus, J., & Reussner, R. H. (2010). Performance modeling in industry: A case study on storage virtualization. In *Proceedings of the 32Nd ACM/IEEE international conference on software engineering – volume 2*, ICSE'10 (pp. 1–10). New York: ACM.
33. Chen, X., Wang, S., Dong, Y., & Wang, X. (2016). *Big data storage architecture design in cloud computing* (pp. 7–14). Singapore: Springer.
34. Hu, H., Wen, Y., Chua, T. S., & Li, X. (2014). Toward scalable systems for big data analytics: A technology tutorial. *IEEE Access, 2*, 652–687.
35. The Hadoop Distributed File System. (2017). http://www.aosabook.org/en/hdfs.html. [Online; Accessed 29-Aug-2017].
36. Kosmos distributed file system (KFS). (2017). http://kosmosfs.sourceforge.net/. [Online; Accessed 29-Aug-2017].

37. NoSQL. (2017). http://nosql-database.org/. [Online; Accessed 29-Aug-2017].
38. BigTable. (2017). https://cloud.google.com/Bigtable/. [Online; Accessed 29-Aug-2017].
39. MongoDB. (2017). https://www.mongodb.com/. [Online; Accessed 29-Aug-2017].
40. Pino, T., Choudhury, S., & Al-Turjman, F. (2018). Dominating set algorithms for wireless sensor networks survivability. *IEEE Access, 6*(1), 17527–17532.
41. Dryad. (2017). https://www.microsoft.com/en-us/research/project/dryad/. [Online; Accessed 29-Aug-2017].
42. Zhang, Z., Cherkasova, L., Verma, A., & Loo, B. T. (2012). Automated profiling and resource management of pig programs for meeting service level objectives. In *Proceedings of the 9th international conference on autonomic computing*, ICAC'12 (pp. 53–62), New York. ACM.
43. Sandholm, T., & Lai, K. (2009). Mapreduce optimization using regulated dynamic prioritization. In *Proceedings of the eleventh international joint conference on measurement and modeling of computer systems*, SIGMETRICS'09 (pp. 299–310), New York. ACM.
44. Graysort benchmark. (2017). http://sortbenchmark.org. [Online; Accessed 29-Aug-2017].
45. Terabyte sort on Apache Hadoop. (2017). http://sortbenchmark.org/Yahoo-Hadoop.pdf. [Online; Accessed 29-Aug-2017].
46. Baru, C., Bhandarkar, M., Nambiar, R., Poess, M., & Rabl, T. (2013). *Setting the direction for big data benchmark standards* (pp. 197–208). Berlin/Heidelberg: Springer.
47. Cooper, B. F., Silberstein, A., Tam, E., Ramakrishnan, R., & Sears, R. (2010). Benchmarking cloud serving systems with YCSB. In *Proceedings of the 1st ACM symposium on cloud computing*, SoCC'10 (pp. 143–154), New York. ACM.
48. Ghazal, A., Rabl, T., Hu, M., Raab, F., Poess, M., Crolotte, A., & Jacobsen, H.-A. (2013). Bigbench: Towards an industry standard benchmark for big data analytics. In *Proceedings of the 2013 ACM SIGMOD international conference on management of data*, SIGMOD'13 (pp. 1197–1208), New York. ACM.
49. Big Data Software. (2017). http://www.kdnuggets.com. [Online; Accessed 29-Aug-2017].
50. The R Project for Statistical Computing. (2017). https://www.r-project.org/. [Online; Accessed 29-Aug-2017].
51. RapidMiner. (2017). https://RapidMiner.com/. [Online; Accessed 29-Aug-2017].
52. KNMINE. (2017). https://www.knime.org/. [Online; Accessed 29-Aug-2017].
53. WEKA. (2017). http://www.cs.waikato.ac.nz/ml/weka/. [Online; Accessed 29-Aug-2017].

Blockchain and Internet of Things-Based Technologies for Intelligent Water Management System

Eustace M. Dogo, Abdulazeez Femi Salami, Nnamdi I. Nwulu, and Clinton O. Aigbavboa

1 Introduction

Water is a critical commodity that connects every aspect of the day-to-day running of cities and communities with direct sociopolitical and economic implications. With projected population explosion, rapid urbanization and climate change induced by the volatility of weather patterns across the globe calls for efficient ways of conserving, utilizing, and managing of the dwindling water resource [53], particularly in emerging economies such as the African region. It is estimated that 70% of world population will live in cities in water stressed regions by the year

E. M. Dogo (✉)
Department of Electrical and Electronic Engineering Science and Institute for Intelligent Systems, Faculty of Engineering and the Built Environment, University of Johannesburg, Johannesburg, South Africa
e-mail: eustaced@uj.ac.za

A. F. Salami
Computer Engineering Department, Faculty of Engineering and Technology, University of Ilorin, Ilorin, Nigeria
e-mail: salami.af@unilorin.edu.ng

N. I. Nwulu
Department of Electrical and Electronic Engineering Science, Faculty of Engineering and the Built Environment, University of Johannesburg, Johannesburg, South Africa
e-mail: nnwulu@uj.ac.za

C. O. Aigbavboa
Department of Construction Management and Quantity Surveying, Faculty of Engineering and the Built Environment, University of Johannesburg, Johannesburg, South Africa
e-mail: caigbavboa@uj.ac.za

F. Al-Turjman (ed.), *Artificial Intelligence in IoT*,
Transactions on Computational Science and Computational Intelligence,
https://doi.org/10.1007/978-3-030-04110-6_7

2025 and the current centralized piped infrastructure being relied upon by water utilities will be inadequate. According to the United Nations (UN) world population prospects, 2017 revision, Europe's population will shrink by 26 million between 2017 and 2050, whereas Africa's population will grow by 1.27 billion over the same period, which makes Europe's challenge tiny in comparison to Africa's.

Internet of Things (IoT) stands out as the core technology and solution to address water conservation and management issues, through digitization and building in intelligence in the water management system, to address water insecurity and attaining the 2030 United Nations' target for the Sustainable Development Goal (SDG 6) on clean water and sanitation [64].

IoT has dramatically changed the way we interact with each other and the environment and provides us with a better way of understanding what's going on around us. IoT is also playing a central role in making our cities and infrastructure more intelligent. It is projected by Cisco Internet Business Solutions Group (IBSG) that the number of interconnected IoT devices will reach 50 billion by 2020, just two (2) years from now, and with an even higher prediction by ARM, a semiconductor and software design company, of 1 trillion IoT devices by 2035. Despite all the promises IoT provides, a key concern about IoT technology is centered around security and privacy of the data continuously being generated by these large-scale sensing devices [25, 55], heterogeneity of devices, scalability, power efficiency and interoperability, and standardization of network communication protocols [51]. Many researchers have proposed several ways of tackling these IoT challenges such as in [46, 63], for real-time IoT [10] and in Industrial Internet of Things (IIoT) industry application scenario [2]. However, there are still unresolved issues and challenges especially with security and privacy owing to the decentralized topology of IoT devices that are constantly prone to physical and cyberattacks due to the sensitive information generated and transmitted by these smart and resource-constrained devices [18].

In [46], researchers identified the following specific security threats to IoT sensors deployed in an intelligent water management system (IWMS):

- Physical attack on sensor devices deployed in remote field locations
- Network hacking and jamming, device cloning, and eavesdropping
- Challenge with securely updating firmware of sensors already deployed in the field
- Issue with secure and trusted communication between devices

Blockchain, a peer-to-peer (P2P) distributed ledger technology (DLT), is emerging as a viable option to improve on the security and other technical challenges of IoT by integrating it with IoT devices to provide a secure network communication platform in a decentralized and distributed architectural arrangement, away from the traditional centralized databases in a cloud-centric design that most IoT has been built upon [15, 19, 59].

The chapter seeks to explore the impact of the combined usage of blockchain and IoT technologies on intelligent water management system (IWMS) through an African perspective. This chapter also summarizes security solutions specifically designed for IWMS and discusses some use case scenarios that could benefit from the hybridization of these two technologies. Although there are works in literature on blockchain-IoT integration [4, 12, 25, 33, 35], to the best of our knowledge, there is currently no study of its applicability in the context of IWMS with an African view and perspective.

2 Overview of Background Concepts

In this section, an overview of the key concepts and technologies is given as it relates to intelligent water management system (IWMS).

2.1 Internet of Things

2.1.1 Technical Definition, Features, and Architecture

IoT is a broad and vast area of research encompassing numerous sets of intertwined ideas from different perspectives. It represents the coming together of several enabling technologies as IP-based networking, wireless communication, miniaturization as envisioned in Moore's law, data analytics, and cloud computing [52], comprising people, machine, and information and having a direct and unprecedented impact on the society in so many ways. It is in light of this that it becomes challenging to give a holistic definition of IoT due to its complex characteristics, different perspectives, and deployment scale. Two definitions of IoT are adopted for the purpose of this chapter, which are based on deployment scale as proffered in [42] as follows:

- Small-scaled IoT deployment: "An IoT is a network that connects uniquely identifiable "Things" to the Internet. The "Things" have sensing/actuation and potential programmability capabilities. Through the exploitation of unique identification and sensing, information about the "Thing" can be collected and the state of the 'Thing' can be changed from anywhere, anytime, by anything."
- Large-scaled IoT deployment: "Internet of Things envisions a self-configuring, adaptive, complex network that interconnects 'things' to the Internet through the use of standard communication protocols. The interconnected things have physical or virtual representation in the digital world, sensing/actuation capability, a programmability feature and are uniquely identifiable. The representation contains information including the thing's identity, status, location or any other business, social or privately relevant information. The things offer services, with or without human intervention, through the exploitation of unique identification, data capture and communication, and actuation capability. The service is exploited through the use of intelligent interfaces and is made available anywhere, anytime, and for anything taking security into consideration."

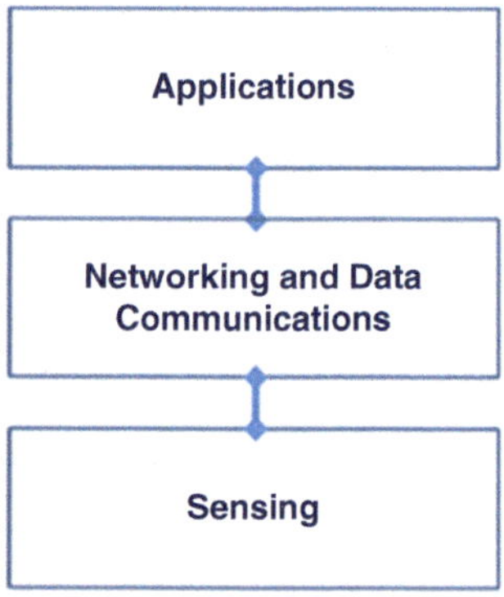

Fig. 1 Simplified IoT functional architecture

IoT has the following characteristics and features:

- Interconnection of devices and between devices and the Internet
- Devices have unique identifier, such as IP and MAC addresses
- Sensing and actuation capabilities
- Inbuilt intelligence
- Self-configurability and programmability

Since smart devices are broad and diverse such as sensors, RFIDs, smart phone, and people, it becomes more glaring and challenging to have a single and common IoT reference model that is applicable to all domains and scenarios. However, a common and simplified three-layered IoT functional reference architecture comprising of physical sensing devices, network with data communication, and application layers, as depicted in Fig. 1, is adopted in many literatures. It is worth mentioning the study by researchers in [51], where an IoT-based reference architecture is proposed for various smart water management processes by using a model developed for a project to establish water management standardization applied in irrigation, known as MEGA, and adopting an existing standard from the manufacturing and logistic domains for the control of processes, known as Object Linking and Embedding for Process Control Unified Architecture (OPC UA). The aim of this study was to define a feasible and all-encompassing industrial standard for water management processes.

2.1.2 IoT Technical Issues and Challenges

IoT can be beneficial to water management in the following ways: water leakage detection, water quality and security monitoring, transparency on consumption and awards of water contracts, predictive maintenance on infrastructure, and more efficient water management system through smart metering system. Despite these benefits that IoT offers, there are still some issues with IoT implementation mainly bordering around security and privacy, regulatory and legal concerns, standards and interoperability, and developmental issues in emerging economies [52]. These mentioned issues are briefly explained:

- Security and Privacy Issues: With rapid entrenchment of IoT in virtually all spheres of our daily lives, concerns are being raised as to degree of security and privacy of the data generated, transmitted, and analyzed over the Internet by these sensing devices, as well as the access rights of this sensitive information. Commonly known cyberattacks on IoT devices include DDoS attack, hacking, data theft, and remote hijacking.
- Interoperability and Standardization: This issue arises due to a combination of multiple technologies that make up the IoT ecosystem, such as sensor technology, wireless communication, embedded system, cloud computing and virtualization, and IoT hardware design considerations.
- Regulatory, Legal, and Right Issues: This issue is related to the way data continuously generated by IoT devices is being utilized, as well as who have access to the data, especially if the data is stored and analyzed across international borders with different jurisdictions from where the data is being generated, raising concerns about user rights, information misuse, and legal liabilities.
- Emerging Economy and Developmental Issues: The IoT hold promise to improve on the economic, political, and social status of emerging and developing nations, just as with the developed ICT economies. However, to realize the full potentials of IoT, the unique African challenges must be taken into consideration.

Readers are referred to [28, 42, 52] and for detailed discussion on technical IoT concept, challenges, standardization, and economy prospects.

2.2 *Blockchain*

2.2.1 Technical Definition and Features

Blockchain is a peer-to-peer (P2P) distributed ledger technology (DLT) for transparent transaction devoid of a trusted intermediary that leverages on the Internet, originally developed for cryptocurrency virtual currency transactions. The concept and idea were first proposed in [22] on time-stamping of a digital document, before it was later called blockchain in [44]. *Blockchain is defined as an appendable immutable universally distributed open ledger* [36]. The key elements of this definition rest with the keywords: *appendable* means can add to the ledger, *immutable* means nothing can be deleted or altered from the ledger, and *universally distributed* means equal accessibility of everyone in the network to the same copy of the ledger each time information is updated to ensure validity of all transactions called blocks. This makes a blockchain trustworthy and an open ledger database where all transactions (blocks) are recorded in a clear, shared, and transparent manner securely linked together using cryptography. We refer readers to [6, 14, 44] for details on blockchain technology.

Currently, many blockchain start-ups are springing up in Africa [7] as shown in Table 1, mostly in the banking and financial sector, rightly so, because of its origin from the cryptocurrency virtual currency world. We however envision these and other new start-ups venturing into the water management ecosystem.

Table 1 Blockchain technology start-ups in Africa

Start-ups	Network consensus platforms	Country	Application domain
Blockchain Academy	Bitcoin/tellar/Ethereum/ IPFS/Hybrid systems	South Africa	Education/social engagement
Satoshicentre		Botswana	Education/training/ social projects
Wala		Uganda	Finance
BitPesa		Kenya	Finance/forex transactions
BitGive		US based (But with partnership operations in Africa)	NGO/charitable organiza-tion/philanthropy
SureRemit		Nigeria	Finance/noncash remittance
Custos Media Technologies		South Africa	Media and music industry piracy
Kobocoin		Nigeria	Finance/payment system
Cryptogene		Nigeria	Education/training
BitMari		Zimbabwe	Finance/forex transactions
ChamaPesa		Kenya	Library/bookkeeping system
NairaEx		Nigeria	Finance/exchange and remittance
Bankymoon		South Africa	Energy and utilities payments/smart grid/consultancy
BitFinance		Zimbabwe	Finance/noncash remittance
The Sun Exchange		South Africa	Solar energy marketplace connect platform
Bitland		Ghana	Land and properties registry
GeoPay		South Africa	Finance/forex remittance
OTLW		Kenya	Online educational system

2.2.2 Components of Blockchain

- Blockchain structure
- Hash function and encryption
- Peer-to-peer (P2P) distributed network
- Consensus protocols
- Mining

2.2.3 Blockchain Stakeholders

The parties, groups, and organizations with interest and concern in this innovative technology are blockchain infrastructure maintainers, external auditors, regulators and law enforcement of blockchain activities and operation, blockchain users or miners, and other pertinent stakeholders.

2.2.4 Benefits of Blockchain to IoT IWMS

The following are well-known benefits which blockchain and IoT convergence promises [17, 37, 59]:

- IoT autonomous transactions using smart contracts in a secure manner among participating devices
- Allows for shared transactions in private blockchain ledgers among parties involved in transactions
- Visibility and transparency in water trading
- Identity and asset management
- Cost reduction through blockchain model as against cloud model
- Tamperproof data
- Open and transparent transactions
- Elimination of intermediaries to consummate transactions
- Improved trust
- Availability of historical records of transactions on multiple IoT smart devices
- Cost reduction linked to deployment and maintenance of Internet infrastructure
- Faster rate of consummating transactions
- Enabling distributed file sharing

2.2.5 Comparison of IoT with Blockchain

Table 2 summarizes the comparison between IoT and blockchain.

Table 2 Comparison between IoT and blockchain [25]

Features	IoT	Blockchain
Security	Low	High
Computational cost	Low	High
Bandwidth	Low consumption	High consumption
Decentralization	No	Yes
Latency	Low	High block mining time
Scalability of nodes	Scales well	Scales poorly

Table 3 IoT challenges and potential blockchain solutions

IoT challenges	Potential blockchain solution
Cost, capacity, and exponential growth of IoT devices constraints	Allows for autonomous, decentralized, and secure communication among devices through smart contracts
Vulnerabilities in IoT architecture, such as hacking, DDoS, and data theft	Cryptographically secured and verified communication/transactions between participating devices
Impact of cloud server downtime and inaccessibility of services owing to several causes, since IoT rely on this arrangement	Elimination of single point of failure, as records are decentralized on numerous devices within the network
Trust and manipulation of data residing in a central location	Decentralization, immutability, and auditability of all transactions

2.2.6 Blockchain-IoT Integration Challenges and Solutions

A summary of IoT challenges and potential blockchain solutions based on a study conducted in [37] is presented in Table 3.

2.3 Intelligent Water Management Network and System

Increasing global population, unpredicted weather patterns, increasing demand for safe water supply by consumers, water stress, and aging water infrastructure call for an innovative and sustainable means of safe and secure water supply for domestic and other usage by water utilities. One key challenge is the loss of water in the distribution network due to pipe bursts, technically known as non-revenue water (NRW), which has direct effect on increased power consumption needed to pump water on the water utilities, financial loss [56], as well as a heightened health risk on consumers. This calls for the use of sustainable technologies for intelligent water management in order to mitigate and proactively monitor the entire water system. Intelligent water system is the harmonious use of water among people, industry, and the environment in a sustainable, safe, and secured water environment and supply service and contributing toward low energy consumption and carbon

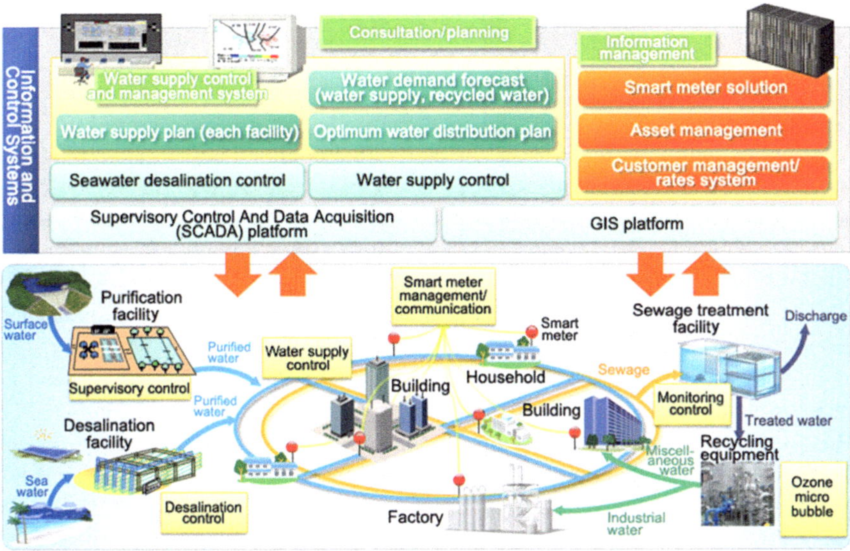

Fig. 2 Intelligent water system conceptual framework [26]

emission [61]. Smart water network is an *integrated set of products, solutions and systems that enable utilities to remotely and continuously monitor and diagnose problems, pre-emptively prioritize and manage maintenance issues and use data-driven insights to remotely control and optimize all aspects of the water distribution network,* as well as *comply transparently and confidently with regulatory and policy requirements on water quality and conservation* and *provide water customers with the information and tools they need to make informed choices about their behaviors and water usage patterns* [56]. The overall objective of smart water solutions includes smart measuring and monitoring across the water distribution, enhanced security, better analysis of the generated data, and enhanced revenue and efficiency [56]. Figure 2 depicts Hitachi's [26] conceptual intelligent water management system which integrates smart supervisory control, instrumentation, information management, and water treatment system, taking into consideration the balance between people and environment in the water environment, ensuring that demand for safe and stable water supply is met in conjunction with concrete environmental conservation strategies for the entire water cycle – water resource cycle, water use, and recycling. In addition to this, practical efforts are as well made to ensure maintenance management of the utility's infrastructure and overall operating efficiency using information and communications technology (ICT).

3 Blockchain-IoT Solutions for Smart Water Management

3.1 Overview of Global Perspective

As a result of the recent increase in cases of droughts, floods, hurricanes, and tsunamis, governmental agencies and private bodies around the world are now compelled to devise smart measures and build sustainable infrastructures for ensuring water security and management [21, 41]. This goal of implementing pragmatic solutions for securing the quality and quantity of water in rural regions and urban areas is still a technologically challenging task which has been successfully tackled in select developed ICT economies like Australia and Singapore [38, 43, 50].

In Australia, the South East Queensland (SEQ) water grid was established as an integrated system for managing and securing urban water supply [38, 43, 51]. This smart integrated system consists of water demand forecasting platforms, water reservoirs and connected dams, water treatment installations, desalination plants, water pumping stations, duplex pipelines for two-way water movement, and, most importantly, a wide range of water sources through recycling, desalination, and precipitation/rainwater [11, 38, 43]. The key benefit of incorporating alternative sources of water supply in the SEQ water grid is to curb scarcity and achieve climate resiliency and critical crisis prevention [11, 51, 57].

In Singapore, the water supply network (WSN) of the Public Utilities Board (PUB) was set up to contrive a self-sufficient network for water management through a holistic process that involves water collection, reclamation, production, and distribution [1, 38, 48]. This holistic process was implemented through the nationwide deployment of smart sensors for real-time monitoring and analytic tools for decision support system in order to achieve an effective and sustainable water supply network management [38, 43]. The operational focus of this smart water grid are water quality monitoring (real-time monitoring, event prediction, and demand forecasting), water conservation (recycling, prevention of wastage, and optimum usage), automated meter reading (real-time water utility information/updates), leak management (leaks/faults detection and preventive maintenance), and asset/infrastructure management (risk management platform development, data mining, and database integration) [1, 43]. WSN also employs a wide range of water sources such as desalinated brine/seawater, harvested rainwater, reclaimed water, and stormwater. This smart water grid consists of water reclamation plants, desalination plants, duplex pipelines, water reservoirs, waste management system, and other core integral infrastructures [1, 38].

3.2 African Perspective

The pace of smart water grid development and intelligent water management in Africa is still very slow due to endemic issues associated with limited access to clean water and lack of reliable water supply/sources [8, 30, 32]. In addition to this, rapid population growth coupled with climate change and other catalysts of water insecurity such as reckless industrial practices, inconsistent water policies, loose water quality control, feeble infrastructure, intentional pipeline damages, poor maintenance culture, high operational costs, huge water utility bills/charges, wanton domestic water wastage, and frequent uncontrolled water losses has subjected many Africans living in urban and rural areas to perennial water scarcity and other severe water-stressed conditions [4, 20, 23, 39, 40]. The implication of these existing challenges is that about 50% of the water supply is inevitably lost before reaching the domestic and industrial consumers in Africa. Nonetheless, there are genuine efforts recently geared toward intelligent water management in Africa [8, 58].

In Kenya, HydroIQ is an indigenous innovation from *Hydrologistics Africa* (Techstars Company) [8]. This solution operates as a water virtual network operator (WNVO) by incorporating smart/automated metering device into the existing traditional water supply network [47]. The WNVO relies on analytic tools, IoT, and payment automation [8, 47]. Consequently, this smart device integration transforms the old traditional water infrastructure into smart water grids by infusing intelligence into water distribution and allowing real-time pipeline condition monitoring for leaks/faults, water pressure, and water quality [8, 47]. In addition to this, by setting up and installing this technology at households, it makes it possible to monitor and control domestic water consumption, thereby allowing local consumers to pay for only the consumed quantity on a pay-as-you-go (PAYG) basis using mobile payment or any other preferred electronic payment channel [8, 47].

In Kyuso, a largely rural district of Kenya, Oxford University researchers implemented the idea of harnessing the pervasive and penetrative powers of mobile networks for indicating when hand pumps are dysfunctional or nonfunctional [31]. These smart hand pumps rely on pump handle movement, mobile data transmitter, water flow measurement, real-time hourly water usage reporting, database processes, information processing, pump location estimator, backbone/central server, and online pump condition alerts [31].

Another innovative solution is the *MajiData* developed by Kenya's Ministry of Water and Irrigation in conjunction with Water Services Trust Fund and in cooperation with the German Development Bank (KfW), German Cooperation for International Cooperation (GIZ), United Nations Human Settlements Programme (UN-Habitat), and Google [31]. Aided by satellite imagery, this online database service provides vital data on slums and low-income areas for water boards and water service providers in order to systematically plan and implement effective water distribution and sanitation schemes for the slums and urban low-income areas [31]. Another case worth mentioning is the technological cooperation and intervention of Culligan (United States Water Treatment Company) in Kigali, Rwanda, to build a smart water grid for the area [49].

3.3 Blockchain-IoT Conceptual Framework

Cloud computing has been an enabling technology for IoT for storage, processing, and analysis of data generated by IoT sensing devices in real time. However, with the proliferation of IoT devices projected to reach over 50 billion mark in the near future, coupled with the assumptions of good Internet connectivity, high bandwidth, and low latency upon which cloud computing concepts have been built on, this cloud-centered architectural arrangement is no longer going to be optimally feasible. Other issues such as trust and transparency are also legitimate concerns raised by cloud computing stakeholders. Even though cloud computing over the years has evolved into new and complimentary concepts such as edge, fog, mist, and dew computing to enhance the capabilities of the cloud computing, most of these inherent challenges with cloud computing persist. Hence, the integration of blockchain and IoT is a promising synergy that could enhance the potential of IoT by providing security, trust, fidelity, and confidentiality in information sharing since the data will remain immutable, appendable in a universally distributed open ledger. Decentralization, scalability, identification of nodes in a trustworthy manner, autonomy, reliability, security, and secured deployment of firmware into devices are some of the benefits identified in [50] that the integration of blockchain-IoT promises. For blockchain-IoT integration in IWMS, we suggest a hybrid approach integration proposed in [50], where cloud computing together with its evolving new concepts like edge, fog, and mist computing could be incorporated into the design to play a complimentary role, as depicted in Fig. 3.

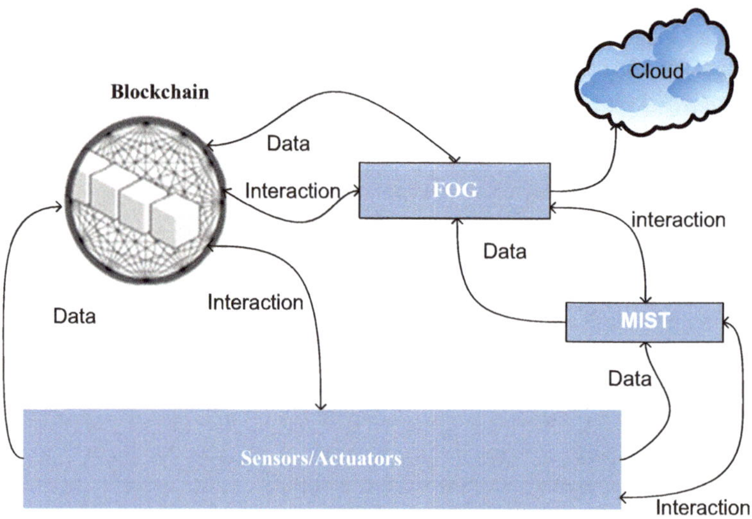

Fig. 3 Conceptual hybrid blockchain-IoT integration (adapted from) [50]

4 Potential Use Cases in Intelligent Water Management

4.1 Stormwater Management

Technical Description and Existing Challenges: With respect to stormwater management, most African cities still adopt ineffective techniques, and as a result, the accumulated runoff of unabsorbed water due to impermeable surfaces lead to pollution, flooding, and other effects deleterious to the environment [16, 37]. In technologically advanced countries, an innovative blend of green infrastructure, gray infrastructure, and smart IoT solutions are employed, deployed, and strategically positioned to monitor stormwater, runoff volumes, peak flows, and short-term and long-range weather conditions in order to take preventive, effective, and sustainable measures for water recycling/reuse, water conservation, and environmental protection [16, 29]. In this research context, gray infrastructure is the stock of engineered facilities and equipment such as water treatment plants, sewage system, and piped drainage system [16, 54]. On the other hand, green infrastructures are the eco-friendly structures like stormwater planters, urban tree canopies, constructed wetlands, green roofs, rainwater harvester, bio-retention facilities, downspout pipes, permeable pavements, detention ponds, retention basins, and land conservation schemes [16, 62]. The smart IoT infrastructures are usually autonomous devices with inbuilt intelligence for event sensing, condition monitoring, data caching/storage, status reporting, and decentralized processing and, most importantly, equipped with seamless and real-time accessibility [3, 27, 45]. In these developed ICT economies, stormwater is considered as a vital resource, and the objectives of blending this tripartite infrastructure are to achieve effective stormwater channeling, storage for evacuation, treatment of water discharge, flow regulation, reintegration, and reuse of stormwater [9, 45].

Current State and Opportunities: A pertinent use case scenario is that of Dogondoutchi city in Niger Republic where land reforms, hillside retention systems (half-moon terraces), desert reforestation, and infiltration systems (filter dykes) were used to control stormwater runoff and quell water stagnation issues in order to revitalize aquaculture, local farming, and irrigation [32]. Another relevant use case scenario is a project funded by the Danish International Development Agency (DANIDA) which is a technological partnership between the Department of Geosciences and Natural Resource Management (IGN), University of Copenhagen, Denmark, Institute of Human Settlements Studies (IHSS) at Ardhi University, Tanzania, and Ethiopian Institute of Architecture, Building Construction and City Development (EiABC) at Addis Ababa University, Ethiopia [62]. The initiative of this project is to introduce and implement the concept of landscape-based stormwater management (LSM) to the African cities of Dar es Salaam and Addis Ababa [62]. LSM entails adopting a robust green infrastructure coupled with adequate storage capacity for the retention, recycling, and reuse of accumulated stormwater runoff during rainy seasons in order to ensure sustainable and intelligent water management [62]. The practical implementation of the LSM employs local retention structures, infiltration systems, land conservation schemes, and other green infrastructures pertinent to these African cities [62]. The benefit of this eco-friendly approach is that it reduces flood risks, revives urban water supply, can be easily used and managed by local consumers, and, most importantly, provides an alternative to the sewage system which is a gray infrastructure [62]. Another germane use case scenario is that of Atlantis town, Western Cape, South Africa, where stormwater is harvested by infiltrating the runoff into the local aquifer with the aid of two large retention ponds [9, 45]. The captured water is injected into a system of boreholes for refinement, reintegration, and reuse in a process popularly known as aquifer recharge management [9, 45]. This intelligent water management technique has ensured sustainable and steady water supply in the region for many years [9, 45].

Recommendation – Public-Private Partnerships and Strong Linkages Between Rural Authorities and Urban Stakeholders: There is the urgent need for respective governments of different African countries to establish runoff monitoring and control unit in strategic rural regions in order to reduce, reuse, and recycle stormwater and wastewater. This can only be achieved through the synergy of government agencies, corporate bodies, rural engineering departments, environmentalists, land use experts, sustainability specialists, industrial technologists, and other stakeholders.

Access to Up-to-Date Technologies, Management Knowledge, Knowhow, Technical Information, Operational Data, and Vital Statistics: Local water authorities and agencies in different African countries need to genuinely and judiciously invest in green infrastructure, technical knowledge, reliable meteorological data collection, storage and analysis, technology management, analytic and forecasting tools, blockchain-IoT trends, sustainable water practices, and other indispensable innovations needed for stormwater management. This will reinforce and better equip the local water authorities to adopt predictive, proactive, and preventive measures for intelligent water management.

4.2 Water Quality Monitoring and Reporting

Technical Description and Existing Challenges: In most African cities, water quality monitoring and smart water metering is still lacking, and most water contamination cases are often detected and reported by the local consumers who have been already exposed to the hazard [5, 24]. It must be highlighted that water contamination occurs more in the water distribution system than in the treatment plants [5, 32]. Contamination in the water distribution system are as a result of colorization, microbial intrusion, disinfectant expiration, biofilms, water staleness, pressure differentials, biochemical abnormalities, industrial accidents, treatment errors, storage wears, pipe stress and bursts, untidy pipe works, pipe corrosion, release of pollutants/toxicants, and acts of sabotage [13, 60]. Apart from this, one of the biggest challenges to optimum water resource management, equitable water distribution, and reliable water access is the issue of losses attached to non-revenue water [5, 40]. Non-revenue water is as a result of unbilled authorized consumption (from unbilled un-metered consumption and unbilled metered consumption), physical losses (from storage leaks, transmission leakages, storage tanks overflow, distribution mains supply leakages, and metering service connection leaks), and commercial losses (from end-user meter inaccuracies, fraudulent manipulation, unauthorized consumption, systemic bugs, and data processing/handling errors) [5, 40].

In developed countries, water quality monitoring in smart water grids is usually achieved by deployed multi-contaminant sensors and biosensors through physio-chemical measurement (to detect pH, optical properties, temperature, turbidity, flow, electrical conductivity, chlorine content, oxidation reduction potential (ORP)) and/or microbial measurement (to detect toxicity) [3, 13]. These parameters are sensed in real time, measured with online instrumentation, and connected to microcontrollers for processing and analyzing the water quality data through event detection and forecasting systems [13, 58]. Local water authorities are then promptly alerted on the location and severity of the detected anomalies in order to take proper preventive actions [27, 60]. The real-time sensed data is also useful for continuous water sampling and hydraulic model calibration which is a replacement to the costly, monotonous, and tedious traditional calibration process [43, 48]. Specifically, smart water meters are used to effectively curb non-revenue water costs, control water wastage, and address pressure differentials, leakages, damages, fraud attempts, consumption rate problems, inaccessibility to account and consumption data, service quality issues, charging and billing issues, and uncontrolled water losses through smart pressure management processes, smart step testing, and other intelligent techniques [40, 60]. Smart valves are employed to prevent contaminant intrusion and cut off polluted water from mixing with potable water through contaminant isolation technique and smart flood management schemes [40, 48]. Generally, smart water grids systemically integrate these automated meters, actuators, sensor array, controllers, and analytic tools for intelligent control and sustainable water management in order to ensure high-quality water supply is reliably and efficiently delivered only when and where it is needed [27, 40].

Current State and Opportunities: Power Ledger from Australia is pioneering a number of interesting and innovative blockchain-powered trading platforms for smart water metering data management, especially in the city of Freemantle [60]. With respect to water quality monitoring and smart water metering, the African market is growing with a good prospect due to the presence of important players coming up with innovative solutions to serve the purpose of intelligent water management. Some of these players are Junaco Trading from Kenya, Aqua-loc from South Africa, Maji Milele Limited from Kenya, Honeywell Elster in South Africa, Metron-Farnier from the United States of America, Amanzi Meters from South Africa, Ideal Prepaid from South Africa, Lesira-Teq Smart Metering from South Africa, Krohne in South Africa, PEC Africa from South Africa, Kamstrup from South Africa, and other technology players [5, 34]. For example, iMvubu from Amanzi Meters employs AMR and GSM for remote meter reading and provides equal treatment to all local consumers by restricting water consumption to a predetermined volume [5].

Generally, smart water metering coupled with water quality monitoring and control is garnering increasing attention and interest in Africa which is forcing local water authorities to take the issue of water conservation, fair/accurate billing, and water management seriously. For instance, water authorities in Nairobi and Nyeri, Kenya, are now issuing meter readers equipped with utility mobile phones which are used in taking, sending, and storing snapshots of correct meter readings for verification and validation purposes [5]. An improvement of this technology was developed which allows local consumers to read meters and cross-validate by simply taking and transmitting an iOS snapshot of the accurate reading to the central utility authorities [5]. In addition to this, some utility authorities in Durban, Dakar, Nairobi, and Nyeri are now incorporating GIS to check city population with respect to water consumption, monitor/report on weather data and climatic conditions, plan service extension, monitor water pipes, and inspect domestic units connected to the water grid and other pertinent information [5].

Recommendation – Developing Multiservice Strategies and Integrated Management Approaches: Water conservation schemes, quality monitoring initiatives, and quality control strategies should not be limited and restricted to the treatment plants, but these techniques should be systematically integrated network-wide (with the aid of IoT sensors, actuators, cognitive systems, and lightweight intelligent codes/programs) in order to achieve end-to-end total quality management for the entire water supply network. This intelligent network will further enhance the performance of the smart water meters by providing a robust, reliable, and resilient water management that can cater for the immediate and future water demands of respective African cities and communities.

4.3 Smart Payment and Contract

Technical Description and Existing Challenges: In Africa, many entrepreneurs operating small- and medium-scale enterprises with the vision of tackling critical water challenges by developing innovative solutions are often restricted as a result of inaccessible loans and standard financing channels from banks due to institutional bureaucracies and stringent collateral terms and conditions [11, 20, 50]. This central financial authority bottleneck slows down innovation process and introduces high risks, verification delays, and currency/country barriers especially in scenarios where there is need for regional cooperation or international collaboration to enable the initial implementation and continued success of the smart water projects [11, 50]. Blockchain is a disruptive and explosive technology that can de-bureaucratize this centralized funding process and facilitate the growth of technological start-ups willing to produce enabling infrastructure, smart facilities, and key IoT solutions that will spur and sustain intelligent water management in Africa [20, 50].

Some developed nations are now developing innovative blockchain applications for water treatment contracts, smart payment, sharing utility data and facility information, water rights trading, and many other interesting applications [11].

Smart payment in this context is done with the aid of a cryptocurrency where every smart water device has its own online "bank account" which is shareable with other devices and/or consumers, and most importantly, micro-transactions enable automated compensation for water consumption [11, 20]. On the other hand, a smart contract is an inbuilt program in the blockchain network for automatically transferring cryptocurrencies between parties (consumers, smart water devices, and other relevant utility units) when certain stipulated conditions are fulfilled [11, 20]. Smart payments and contracts solve the existing inconsistencies and errors associated with the traditional performance evaluation, maintenance, and commercialization of water facilities [11, 50]. By adopting these smart methodologies, business transactions are seamlessly streamlined, costs are minimized, and fraud attempts are effectively checked/prevented [11, 50].

Current State and Opportunities: In Australia, Civic Ledger developed a blockchain-powered P2P platform, Water Ledger that utilizes token management system and smart contract for monitoring water transactions and automatically updating records of state water departments [11]. OriginClear from the United States is developing an application called WaterChain which is a blockchain protocol that employs cryptocurrency and smart contracts to ensure service efficiency and operational transparency in the water treatment industry [11]. AQUAOSO from the United States is architecting blockchain to track ownership of water rights and also integrating smart contracts into existing solutions with the aid of IoT-enabled sensors to monitor water supply chain and quality of service delivery and ensure smart payments [3, 11]. In China, partnership between NW Blockchain Limited and Newater Technology is reached for managing water facilities/plants and industrial wastewater projects through asset-based token sales [11]. In Senegal, Banque Regionale de Marches launched eCFA Franc cryptocurrency for West African Monetary Union in order to drive Internet of Intelligent Things to build and maintain key infrastructure for economic development [24]. In South Africa, Bankymoon launched a blockchain humanitarian platform that allows donors to top up specific schools' meters from any location around the world [11]. Generally, blockchain technology can be integrated into smart meters in order to enable consumers to pay bills with cryptocurrencies and adopt sustainable water practices.

Recommendation – Genuine Interest and Investment in Distributed Computing and Blockchain-Powered IoT Engineering: Respective governments of different African countries need to urgently invest in remote-access data processing, data mining, tele-computing, computational intelligence, big data analytics, and blockchain-based IoT design methodologies. By doing this, transparency will be ensured, and all stakeholders (domestic consumers, industrial users, local water authorities, private investors, rural water managers, conservationists, policymakers, and other pertinent participants in the water sector) will have access to reliable, relevant, and recent data on water quality and quantity in order to formulate policies and make more informed decisions. This will also stem corrupt practices and fraudulent activities especially in scenarios where local water authorities have been bribed to tamper with water quality data and falsify consumption rates and charges. In addition to this, adopting these intelligent distributed computing techniques will catalyze P2P water rights trading enabling local consumers with extra water resources to trade off this excess (or any saleable portion) without involving and relying on the central authority. This smart trading decision is often based on information on market trends, latest weather forecast data, long-range climatic predictions, and other bulk of processed data available and accessible through the Internet and/or mobile devices. This will also boost market fidelity/outlook and open up novel invest opportunities, smart business models, and innovative finance strategies for start-ups and enterprises willing to integrate IoT, AI, and blockchain solutions into their products or designs in order to tackle existing water challenges in various African regions.

5 Conclusion

Blockchain and IoT are still in their developmental stage in Africa. The current trend suggests that blockchain and IoT with artificial intelligence will be dominant technologies in the near future. In this era of the 4IR, Africa must key into the numerous benefits these technologies offer. One such area that could benefit from this synergy is in the water industry; this of course stems from the fact that water connects every aspect of life and it is critical to the social and economic survival of the African region. There is no one-size-fits-all solution to the water and sanitation crisis saddling the region; however, the integration of blockchain and IoT has the power to revolutionize water and sanitation management toward achieving SDG 6 as envisioned by the UN in 2035, through innovative, efficient, scalable solutions, based on these two technologies. This will ensure that African cities are adequately prepared with technologies to respond and address the perennial challenges associated with water crisis, such as in Cape Town of recent and other parts of Africa in a sustainable and efficient way.

Acknowledgments We would like to thank the University of Johannesburg for the funding and affording the resources to complete this work.

References

1. Allen, M. (2012). Case study: A smart water grid in Singapore. *Water Practice and Technology, 7*(4), 1–8.
2. Al-Turjman, F., & Alturjman, S. (2018). Context-sensitive access in industrial internet of things (IIoT) healthcare applications. *IEEE Transactions on Industrial Informatics, 14*(6), 2736–2744.
3. Aquaoso. (2018). The power of IoT AI blockchain & hustle in water. Available: https://aquaoso.com/power-iot-ai-blockchain-hustle-water. Accessed 6 August 2018.
4. Atzori, M. (2016). Blockchain-based architectures for the internet of things: A survey. *SSRN Electronic Journal*. https://doi.org/10.2139/ssrn.2846810.

5. Ayemba, D. (2017). Why Africa needs water metering. Available: https://constructionreviewonline.com/2017/04/africa-needs-water-metering/. Accessed 20 August 2018.
6. Bahga, A., & Madisetti, V. K. (2016). Blockchain platform for industrial internet of things. *Journal of Software Engineering and Applications, 9*(10), 533–546. https://doi.org/10.4236/jsea.2016.910036.
7. Bitcoin Africa. (2017). African's blockchain startups. Available: https://bitcoinafrica.io/category/meet-africas-blockchain-startups/. Accessed 10 April 2018.
8. Bosire, B. (2017). Disrupting the water business. Available: https://www.hydroiq.africa/single-post/2017/12/05/Value-in-every-drop. Accessed 19 August 2018.
9. Carden, K., & Fisher-Jeffes, L. (2017). Stormwater harvesting could help South Africa manage its water shortages. Available: http://theconversation.com/stormwater-harvesting-could-help-south-africa-manage-its-water-shortages-74377. Accessed 17 August 2018.
10. Chen, C-Y., Hasan, M., & Mohan, S. (2017). Securing real-time internet-of-things. Available: http://arxiv.org/abs/1705.08489. Accessed 20 August 2018.
11. Chief Technology Officer. (2018). Hashing out the future of blockchain for the water industry. Available: https://www.originclear.com/pdf/Blockchain-Article.pdf. Accessed 8 August 2018.
12. Christidis, K., & Devetsikiotis, M. (2016). Blockchains and smart contracts for the internet of things. *IEEE Access, 4*, 2292–2303. https://doi.org/10.1109/ACCESS.2016.2566339.
13. Cloete, N. A., Malekian, R., & Nair, L. (2016). Design of smart sensors for real-time water quality monitoring. *IEEE Access, 4*, 3975–3990. https://doi.org/10.1109/ACCESS.2016.2592958.
14. Conte de Leon, D., Stalick, A. Q., Jillepali, A. A., Haney, M. A., & Sheldon, F. T. (2017). Blockchain: Properties and misconceptions. *Asia Pacific Journal of Innovation and Entrepreneurship, 11*(3), 286–300. https://doi.org/10.1108/APJIE-12-2017-034.
15. Corsaro, A. (2016). The cloudy, foggy and misty internet of things - towards fluid IoT architectures. Available: https://icpe2016.spec.org/fileadmin/user_upload/documents/icpe_2016/2016.03.16 Fluid IoT cready.pdf. Accessed 28 August 2018.
16. Cuellar, A., Pallaske, G., & Wuennenberg, L. (2017). Stormwater markets: Concepts and applications. International Institute for Sustainable Development. Available: https://www.iisd.org/sites/default/files/publications/stormwater-markets-concepts-applications.pdf. Accessed 6 August 2018.
17. Dickson, B. (2016). Decentralizing IoT networks through blockchain. Available: http://social.techcrunch.com/2016/06/28/decentralizing-iot-networks-through-blockchain/. Accessed 7 September 2018.
18. Dorri, A., Kanhere, S. S., & Jurdak, R. (2016). Blockchain in internet of things: Challenges and solutions. Available: http://arxiv.org/abs/1608.05187. Accessed 25 August 2018.
19. Fernandez-Carames, T. M., & Fraga-Lamas, P. (2018). A review on the use of blockchain for the internet of things. *IEEE Access, 6*, 32979–33001. https://doi.org/10.1109/ACCESS.2018.2842685.
20. Ferrag, M. A., Derdour, M., Mukherjee, M., Derhab, A., Maglaras, L., & Janicke, H. (2018). Blockchain technologies for the internet of things: Research issues and challenges. Available: http://arxiv.org/abs/1806.09099. Accessed 25 August 2018.
21. Garg, P. (2018). Can blockchain technology help solve the water crisis. Available: http://btcmanager.com/can-blockchain-technology-help-us-solve-the-water-crisis. Accessed 6 August 2018.
22. Haber, S., & Stornetta, W. S. (1991). How to time-stamp a digital document. *Journal of Cryptology, 3*(2). https://doi.org/10.1007/BF00196791.
23. Hachani, A. (2017). IoT as an enabler for smart water management. ITU regional Workshop on Prospects of Smart Water Management (SWM) in Arab Region, Khartoum-Sudan, 12 December 2017.
24. Hall, M. (2017). How to make an Internet of Things work for Africa. Available: http://www.gsbbusinessreview.gbs.uct.ac.za/make-internet-things-work-for-africa. Accessed 6 August 2018.

25. Atlam, H. F., Alenezi, A., Alassafi, M. O., & Wills, G. B. (2018). Blockchain with internet of things: Benefits, challenges, and future directions. *International Journal of Intelligent Systems and Applications, 9*(6), 40–48. https://doi.org/10.5815/ijisa.2018.06.05.
26. Hitachi. (2018). Intelligent Water Systems. Available: http://www.hitachi.com/businesses/infrastructure/product_site/water_environment/intelligent_water/index.html. Accessed 9 September 2018.
27. IIoT World. (2018). Smart water management using Internet of Things technologies. Available www.iiot-world.com/category/connected-industry. Accessed 6 August 2018.
28. ILNAS. (2018). Internet of things (IoT): Technology, economic view and technical standardization. Ministry of the Economy of Luxembourg. Available: https://portail-qualite.public.lu/dam-assets/publications/normalisation/2018/white-paper-iot-july-2018.pdf. Accessed 9 September 2018.
29. Impact Chain Lab. (2017). Managing our water with blockchain. Available: https://medium.com/@impactchainlab/managing-our-water-supply-with-blockchain. Accessed 6 August 2018.
30. IOT Now Magazine. (2018). Case study: 'Intelligence on tap', inside veolia water's smart grid for water services deployment. Available: https://www.iot-now.com/2018/01/19/74838-intelligence-tap-inside-veolia-waters-smart-grid-water-services-deployment/. 3 Sept 2018.
31. ITU News. (2014). Success stories in smart water management. Available: http://itunews.itu.int/en/4852-Success-stories-in-smart-water-management.note.aspx. Accessed 16 August 2018.
32. Jallé, L. C., Désille, D., & Burkhardt, G. (2013). Urban stormwater management in developing countries. Available: http://documents.irevues.inist.fr/bitstream/handle/2042/51221/1A2P07-005LEJ.pdf?sequence=1. 10 Aug 2018.
33. Jesus, E.F., Chicarino, V.R.L., de Albuquerque CV, & Rocha, A. A. D. A (2018) A survey of how to use blockchain to secure internet of things and the stalker attack. *Security and Communication Networks* (Hindawi Limited, 2018, Article ID 967050:1–27, doi:https://doi.org/10.1155/2018/9675050.
34. Kamstrup. (2013). Smart-water-meter-passes-test-in-South-Africa. Available: https://www.kamstrup.com/en-en/news-and-events/news/smart-water-meter-passes-test-in-south-africa. Accessed 19 August 2018.
35. Khan, M. A., & Salah, K. (2018). IoT security: Review, blockchain solutions, and open challenges. *Future Generation Computer Systems, 8*(2), 395–411. https://doi.org/10.1016/j.future.2017.11.022.
36. Kimbel, K. (2018). The secret behind the blockchain technology. Available https://pecb.com/past-webinars/the-secret-behind-the-blockchain-technology. Accessed 14 April 2018.
37. Kshetri, N. (2017). Can blockchain strengthen the internet of things? *IT Professional, 19*(4), 68–72. https://doi.org/10.1109/MITP.2017.3051335.
38. Lee, S. W., Sarp, S., Jeon, D. J., & Kim, J. H. (2015). Smart water grid: The future water management platform. *Desalination and Water Treatment, 55*(2), 339–346. https://doi.org/10.1080/19443994.2014.917887.
39. Lin, Y., Petway, J., Lien, W., & Settele, J. (2018). Blockchain with artificial intelligence to efficiently manage water use under climate change. *Environments, 5*(3), 34. https://doi.org/10.3390/environments5030034.
40. Martyusheva, O. (2014). Smart water grid. Plan B Technical Report, MSc Thesis, Department of Civil and Environmental Engineering, Colorado State University. Available: http://www.engr.colostate.edu/~pierre/ce_old/resume/Theses%20and%20Dissertations/Martyusheva,Olga_PlanB_TechnicalReport.pdf. Accessed 20 March 2018.
41. Mehta, Y. (2016). How the Internet of Things helps in water management. Available: http://iotworm.com/internet-things-helps-water-management-system/. Accessed 8 August 2018.
42. Minerva, R., Biru, A., & Rotondi, D. (2015). Towards a definition of the internet of things (IoT). IEEE internet of things. www.iot.ieee.org.
43. Mutchek, M., & Williams, E. (2014). Moving towards sustainable and resilient smart water grids. *Challenges, 5*(1), 123–137. https://doi.org/10.3390/challe5010123.

44. Nakamoto, S. (2008). Bitcoin: A peer-to-peer electronic cash system. Available: http://bitcoin.org/bitcoin.pdf. Accessed 20 March 2018.
45. Nicolson, A. (2017). Harvesting the storms. Available: http://www.news.uct.ac.za/article/-2017-09-15-harvesting-the-storms. Accessed 19 August 2018.
46. Ntuli, N., & Abu-Mahfouz, A. (2016). A simple security architecture for smart water management system. *Procedia Computer Science, 83*(2016), 1164–1169. https://doi.org/10.1016/j.procs.2016.04.239.
47. Okikie, O. (2018). Full list of 50 African startups exhibiting at AIS Rwanda 2018. Available: https://smepeaks.com/2018/06/01/top-50-african-startups-exhibiting-ais-rwanda/. Accessed 4 September 2018.
48. Public Utilities Board Singapore. (2016). Managing the water distribution network with a smart water grid. *Smart Water, 1*(4), 1–13. https://doi.org/10.1186/s40713-016-0004-4.
49. Quartz Africa. (2017). African countries want to turn their poor, overcrowded urban centers into "smart cities". Available: https://nextbillion.net/news/african-countries-want-to-turn-their-poor-overcrowded-urban-centers-into-smart-cities/. Accessed 18 August 2018.
50. Reyna, A., Martín, C., Chen, J., Soler, E., & Díaz, M. (2018). On blockchain and its integration with IoT. Challenges and opportunities. *Future Generation Computer Systems, 88*, 173–190. https://doi.org/10.1016/j.future.2018.05.046.
51. Robles, T., Alcarria, R., Martin, D., Navarro, M., Calero, R., Iglesias, S., & Lopez, M. (2014). An IoT based reference architecture for smart water management processes. *Journal of Wireless Mobile Networks, Ubiquitous Computing, and Dependable Applications, 6*(1), 4–23.
52. Rose, K., Eldridge, S., & Chapin, L. (2015). The internet of things: An overview – Understanding the issues and challenges of a more connected world. The Internet Society (ISOC). Available: http://www.internetsociety.org/sites/default/files/ISOC-IoT-Overview-20151221-en.pdf. Accessed 6 August 2018.
53. Ryder, G. (2018). How ICTs can ensure the sustainable management of water and sanitation. Available: https://news.itu.int/icts-ensure-sustainable-management-water-sanitation/. Accessed 5 September 2018.
54. Salam, M. (2018). Smart water management using Internet of Things technologies. Available: https://iiot-world.com/connected-industry/smart-water-management-using-internet-of-things-technologies/. Accessed 6 August 2018.
55. Samaila, M. G., Neto, M., Fernandes, D. A. B., Freire, M. M., & Inácio, P. R. M. (2017). Security challenges of internet of things. In J. M. Batalla, G. Mastorakis, C. X. Mavromoustakis, & E. Pallis (Eds.), *Beyond the internet of things: Everything interconnected* (pp. 53–82). Cham: Springer International Publishing.
56. Sensus. (2012). Water 20/20: Bringing smart water networks into focus. Sensus. Available: https://www.swan-forum.com/wp-content/uploads/sites/218/2016/05/sensus_water2020-usweb.pdf. Accessed 9 August 2018.
57. Mendonca, S. F. T. D. O., da Silva Junior, J. F., & de Alencar, F. M. R.. (2017). The Blockchain-based Internet of Things development: Initiatives and challenges. The Twelfth International Conference on Software Engineering Advances (ICSEA 2017), October 8–12, 2017, Athens, Greece.
58. Siemens. (2014). Smart sensors for smart water grids. Available: https://w3.siemens.com/mcms/solution-partner/en/home/industry/Documents/event-2014-10-10/4_Smart_Sensors_for_smart_water_grids.pdf. Accessed 3 September 2018.
59. Snyder, D. (2017). Blockchain technology for the Internet of Things. 42TEK Inc. Available: https://californiaconsultants.org/wp-content/uploads/2016/11/CNSV-1701-SnyderManianGreco.pdf. Accessed 10 August 2018.
60. Stinson, C. (2018). How blockchain AI other emerging technologies could end water insecurity. Available: https://www.greenbiz.com/article/how-blockchain-ai-and-other-emerginging-technologies-could-end-water-insecurity. Accessed 6 August 2018.
61. Tadokoro, H., Onishi, M., Kageyama, K., Kurisu, H., & Takahashi, S. (2011). Smart water management and usage systems for society and environment. *Hitachi Review*, (3), 60, 165.

62. The Stormwater Report. (2014). Danish researchers look at stormwater management in Africa. Available: http://stormwater.wef.org/2014/09/landscape-based-stormwater-management-in-africa/. Accessed 4 September 2018.
63. Tsague, H. D., & Twala, B. (2018). In N. Dey, A. Hassanien, C. Bhatt, A. Ashour, & S. Satapathy (Eds.), *Practical techniques for securing the Internet of Things (IoT) against side channel attacks, in Internet of Things and big data analytics toward next-generation intelligence* (pp. 439–481). Cham: Springer.
64. UN. (2015). Sustainable development goals: 17 Goals to transform our world. Available: https://www.un.org/sustainabledevelopment/. Accessed 6 August 2018.

Digital Forensics for Frame Rate Up-Conversion in Wireless Sensor Network

Wendan Ma and Ran Li

1 Introduction

With the rapid development of the industrial Internet and smart city, more and more wireless sensors are used to collect and exchange multimedia and video data. In applications of various industries, we can see that the transmission and exchange of video data are inevitable. In response to the development of the era, many methods for processing multimedia and video data have been proposed [1–3]. However, the rich temporal redundancy in video data is easy to be exploited by forgers, which makes the digital video face the threat of forgery. Especially in wireless sensor network, the massive data collected by nodes has to actively discard part of video frames during transmission due to limited bandwidth, and then the frame rate up-conversion (FRUC) [4] is implemented to restore video at the receiver. In order to ensure the users' right to know the data integrity and to avoid inappropriate post-processing, it is necessary to design a digital forensics method to identify FRUC forgery [5].

Frame replication (FR) is the simplest FRUC method, which is used by some commercial video editing softwares (e.g., ImTOO, Video Edit Magic, etc.) because of its simplicity and ease of use. Aiming at FR forgery, the forgery traces can be found by analyzing the similarity between adjacent frames. For example, Bian et al. [6] quantify the inter-frame similarity by using structural similarity (SSIM) index [7]. Yang et al. [8] extract the feature of each frame and quantify the inter-frame similarity using the Euclidean distance between features. The periodicity of similarity indexes is a strong evidence of the FR forgery. The high frame rate of video produced by FR often appears flickering and jerkiness, which is caused

W. Ma · R. Li (✉)
School of Computer and Information Technology, Xinyang Normal University, Xinyang, China
e-mail: liran@xynu.edu.cn

F. Al-Turjman (ed.), *Artificial Intelligence in IoT*,
Transactions on Computational Science and Computational Intelligence,
https://doi.org/10.1007/978-3-030-04110-6_8

by ignoring the motion between frames. Therefore, the more advanced method, motion-compensated FRUC (MC-FRUC) [9], is favored by forgers. MC-FRUC interpolates pixels along the motion trajectory, generating a smooth, clear, and comforting video, so it is a popular approach to increase frame rate. However, there is no periodic inter-frame similarity in these high-quality forged videos, which makes the forensics algorithm of detecting FR forgery invalid [6]. Bestagini et al. [10] take the lead in detecting MC-FRUC forgery. First, the suspicious video is down-sampled, and the frame average operator is used to increase the frame rate to obtain the detection video. Then, the variance is calculated frame by frame for the detected video and the suspicious video. Finally, the forgery evidence is obtained by analyzing the periodicity of variance. Bestagini's algorithm achieves certain effects, but the algorithm must know the up-conversion factor. However, this condition cannot be met in most cases. Yang et al. [11–13] also explore the connection between some statistical features and interpolation frame operation, which improve the detection accuracy of MC-FRUC forgery. Since MC-FRUC can be implemented by different technical methods [14–16], the detection accuracy will be greatly degraded when the above MC-FRUC detection algorithm encounters complex high-precision forgery methods (e.g., multi-hypothesis motion estimation [16], post-texture rendering [17]). Especially after the forger conducts compression, de-noising, and other post-processing, the situation will get worse. Therefore, it is a large challenge to ensure the property stability of MC-FRUC forensics algorithm at present.

To improve the detection accuracy of MC-FRUC, this chapter proposes to measure the statistical variation of video data with edge feature. First, the Sobel operator is used to detect the edge of video frames. Then, the edge is quantified to obtain the edge complexity of each frame. Finally, the periodicity of the edge complexity along time axis is detected, and FRUC forgery is automatically identified by hard threshold decision. Experimental results show that the proposed forensics algorithm can effectively identify video sequences forged by different MC-FRUC methods. Especially after de-noising and compression attacks, the proposed algorithm can still ensure high detection accuracy.

The rest of this chapter is organized as follows. Section 2 briefly describes the forensics of FR forgery and MC-FRUC. Section 3 presents the proposed forensics algorithm. Section 4 discusses the experimental results and analysis. We conclude this chapter in Section 5.

2 Background Knowledge

2.1 Forensics of FR Forgery

FR is to increase the frame rate of video by directly copying adjacent frames and inserting in the waiting time. That is, suppose the t-th frame is the interpolated frame, then

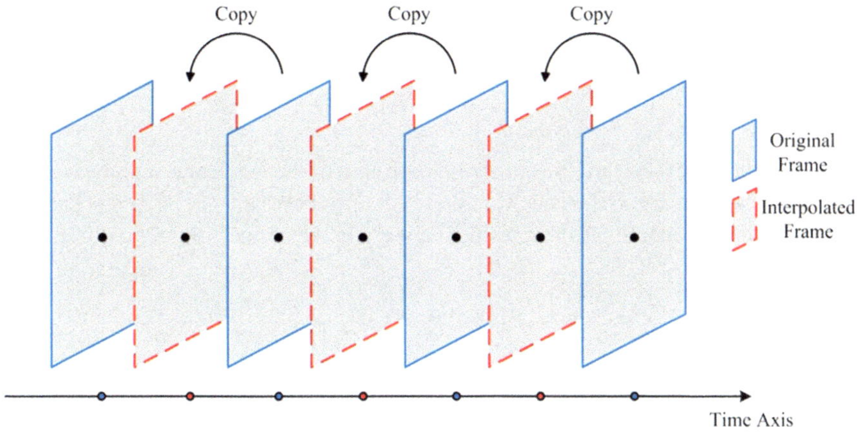

Fig. 1 Example of FR forgery

$$f[t] = f[t+1] \tag{1}$$

where $f[t]$ is the interpolated frame and $f[t+1]$ is the adjacent original frame. As shown in Fig. 1, the frame rate of original video is increased by periodically inserting interpolated frame, which means that there are always video frames consistent with its contents near the interpolated frames. Therefore, it can be proved that the video has FR forgery as long as interpolated frames are found in the neighborhood of a certain time. The existing work adopts two detection methods, namely, residual detection [10] and similarity detection [6]. The abovementioned two detection methods are briefly described below.

1. Residual detection

The residual detection method calculates the inter-frame residual energy of the t-th frame and its adjacent $(t+1)$-th frame as follows:

$$v[t] = \| f[t] - f[t+1] \|_2^2 \tag{2}$$

where $||\cdot||_2$ is l_2 norm. If $f[t]$ is interpolated frame, $v[t]$ is 0; otherwise, $v[t]$ is not 0. Once the interpolated frame appears periodically, the residual energy will periodically generate the value 0. Therefore, as long as the periodic attenuation of the residual energy can be detected to be 0, it can be proved that there is FR forgery. In general, after the frame rate of original video is improved by FR, video coding systems (e.g., H.264, HEVC, etc.) are adopted to implement lossy compression on the forged video in order to reduce the data volume as much as possible, which will lead to a certain error between the compressed video frame and the original video frame. That is,

$$\widehat{f}[t] = f[t] + e[t] \tag{3}$$

$$\widehat{f}[t+1] = f[t+1] + e[t+1] \tag{4}$$

where $\widehat{f}[t]$and$\widehat{f}[t+1]$are the compression frame of the t-th frame and the $(t + 1)$-th frame, respectively. $e[t]$, $e[t + 1]$ is the error term between $\widehat{f}[t]$,$\widehat{f}[t+1]$ and its original frame, respectively. The residual energy between the compressed frames is calculated as follows:

$$\widehat{v}[t] = \left\| \widehat{f}[t] - \widehat{f}[t+1] \right\|_2^2 = v[t] + \| e[t] - e[t+1] \|_2^2 \tag{5}$$

When the forged frame encounters lossy compression, it can be seen from Eq. (5) that the residual energy of the compressed frame includes not only the residual energy of the original frame but also the error energy. The error term will vary randomly over time due to the different quality reductions of each compressed frame. Therefore, the lossy compression will interfere with the periodic variation of residual energy, thereby reducing the detection performance.

2. Similarity detection

The similarity detection method calculates the SSIM value of the t-th frame and its adjacent $(t + 1)$-th frame as follows:

$$s[t] = \mathrm{SSIM}(f[t], f[t+1]) = \frac{(2\mu_t \mu_{t+1} + c_1)\left(2\lambda_{t,t+1} + c_2\right)}{\left(\mu_t^2 + \mu_{t+1}^2 + c_1\right)\left(\lambda_t^2 + \lambda_{t+1}^2 + c_2\right)} \tag{6}$$

where μ_t and μ_{t+1} denote the mean value of $f[t]$ and $f[t + 1]$, respectively. λ_t and λ_{t+1} denote the variance of $f[t]$ and $f[t + 1]$, respectively. $\lambda_{t,t+1}$ denotes the covariance of $f[t]$ and $f[t + 1]$; c_1 and c_2 denote two constants. The SSIM value ranges from 0 to 1, where 1 indicates that the two frames are identical. The smaller the value is, the less similar the two frames are. When the interpolated frame is inserted periodically, the SSIM value will periodically appear as value 1. Therefore, as long as the SSIM value can be detected to periodically increase to 1, it can be proved that there is FR forgery. It is inevitable that noise will be mixed in forged video during transmission. That is,

$$\overline{f}[t] = f[t] + n[t] \tag{7}$$

$$\overline{f}[t+1] = f[t+1] + n[t+1] \tag{8}$$

where $\overline{f}[t]$ and $\overline{f}[t+1]$ denote the noisy frame of the t-th frame and the $(t + 1)$-th frame, respectively. $n[t]$ and $n[t + 1]$ are noise terms, respectively. The SSIM value of the noisy frame is calculated as follows:

$$\overline{s}[t] = \mathrm{SSIM}\left(\overline{f}[t], \overline{f}[t+1]\right) = s[t] + \delta(n[t], n[t+1]) \tag{9}$$

where $\delta(n[t],n[t+1])$ is the interference value caused by the noise terms $n[t]$ and $n[t+1]$. When the forged frame is subjected to noise interference, it can be seen from Eq. (9) that the SSIM value of the noisy frame includes not only the SSIM value of the original frame but also the interference term. When the noise mixed in is too large, it is bound to affect the periodic variation of SSIM value, thus reducing the detection performance.

2.2 MC-FRUC

MC-FRUC is a video post-processing technology, which can realize inter-frame interpolation by predicting inter-frame motion estimation. MC-FRUC can effectively enhance motion continuity and improve the fluency of video. By virtue of abundant original data in the video and high correlation along the motion trajectory, MC-FRUC has attracted wide attention from industry and academia. As shown in Fig. 2, MC-FRUC is mainly composed of motion estimation, motion vector smoothing, motion vector mapping, and motion-compensation interpolation. The first three parts use the reference frames $\boldsymbol{f}_{t-1}$ and $\boldsymbol{f}_{t+1}$ to generate the motion vector field $\boldsymbol{V}_t$ of the current frame $\boldsymbol{f}_t$. The last part is to form the current frame estimation $\widehat{f}_t$ based on $\boldsymbol{V}_t$. First, the motion estimation is used to generate the motion vector field $\boldsymbol{V}_{t-1}$ between $\boldsymbol{f}_{t-1}$ and $\boldsymbol{f}_{t+1}$. Then, the motion vector smoothing corrects the abnormal motion existing in $\boldsymbol{V}_{t-1}$ to obtain a smooth motion vector field $\overline{\mathbf{V}}_{t-1}$, and $\overline{\mathbf{V}}_{t-1}$ is mapped to the current frame motion vector field $\boldsymbol{V}_t$ by motion vector mapping. Finally, according to $\boldsymbol{V}_t$, the matching pixel of any pixel in the current frame can be found in the reference frame, and the current frame is interpolated as follows:

$$\widehat{f}_t(i,j) = \frac{1}{2}\left[f_{t-1}\left(i + V_{t,1}^{(i,j)}, j + V_{t,2}^{(i,j)}\right) + f_{t+1}\left(i - V_{t,1}^{(i,j)}, j - V_{t,2}^{(i,j)}\right)\right] \tag{10}$$

where (i, j) represents the pixel position and $\left(V_{t,1}^{(i,j)}, V_{t,2}^{(i,j)}\right)$ represents the motion vector of $\boldsymbol{V}_t$ at pixel position (i, j). The MC-FRUC interpolation accuracy mainly

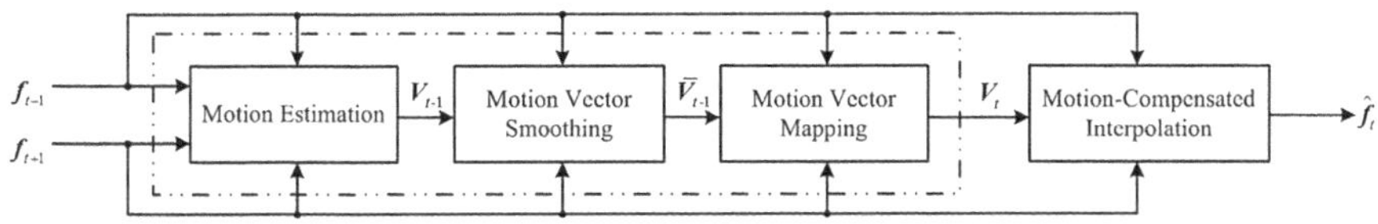

Fig. 2 Framework of MC-FRUC

(a) Method in [14] (b) Method in [15] (c) Method in [16]

Fig. 3 Subjective visual quality comparison of the 98-th frame in *Foreman* sequence interpolated with different MC-FRUC strategies

depends on the accuracy of $\boldsymbol{V}_t$; therefore a lot of works has focused on improving the performance of motion estimation and motion vector smoothing and mapping. For example, in [14], 3-D recursive search (3DRS) is adopted in motion estimation to reduce abnormal motion through implicit smoothing constraints; in [15], motion vector mapping is performed by using bidirectional motion symmetry assumption to avoid the problem of overlapping and holes; in [16], multi-hypothesis motion estimation is adopted to form a more accurate motion vector field by the motion vector fields of different densities. Different strategies are used to realize the motion estimation part, which can provide the motion vector field with different precision levels, so the interpolated frames with different visual qualities can be generated by Eq. (10). As shown in Fig. 3, using the MC-FRUC method in [14–16] to restore the 98-th frame of the *Foreman* sequence in CIF format, it can be seen that the multi-hypothesis motion estimation proposed in [16] provides a better interpolation quality and in [14, 15] proposed algorithm takes second place. Due to the actual demand of "fake real," the high level of MC-FRUC is more popular and widely used in many video applications.

3 Proposed Forensics Algorithm

The flow of proposed forensics algorithm is shown in Fig. 4: First, the original video is subjected to MC-FRUC forgery to generate up-converted video, and attacks may also implement on it such as de-noising and compression; then, the Sobel operator is used to extract the edge feature of each video frame; and finally, analyzing the periodic variation of the edge feature of suspicious video over time to identify whether there is MC-FRUC forgery. The core of the proposed algorithm is the edge feature extraction and periodicity detection, which will be introduced in detail below.

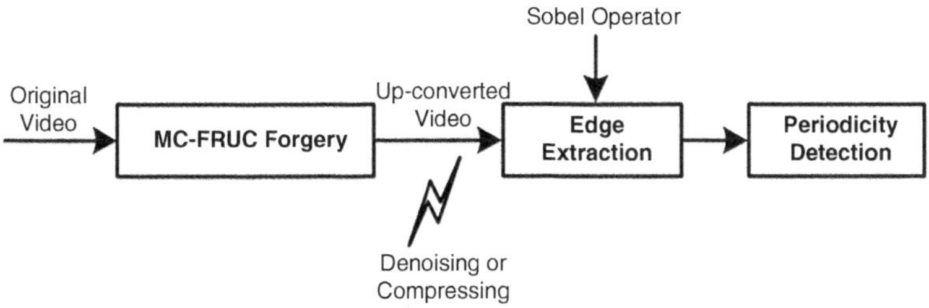

Fig. 4 Flow of the proposed forensics algorithm

3.1 Edge Feature Extraction

Sobel operator is a discrete differential operator that combines Gaussian smoothing and differential derivation, and it is often used to calculate the approximate gradient of the image [18]. Image edge represents the jump process of pixels, and the gradient is the way to measure the degree of jump. The large gradient value indicates the significant improvement of pixel value and reflects the distribution of edge features in the image. Let the original video sequence to be composed of L video frames with size $M \times N$. For the t-th frame $\boldsymbol{f}_t$, using the Sobel operator to realize the edge detection in horizontal and vertical direction as follows:

$$\boldsymbol{g}_h = \begin{bmatrix} -1 & 0 & +1 \\ -2 & 0 & +2 \\ -1 & 0 & +1 \end{bmatrix} * \boldsymbol{f}_t \tag{11}$$

$$\boldsymbol{g}_v = \begin{bmatrix} -1 & -2 & -1 \\ 0 & 0 & 0 \\ +1 & +2 & +1 \end{bmatrix} * \boldsymbol{f}_t \tag{12}$$

where * denotes convolution operation. Geometrical average is calculated pixel by pixel for edge pixel values in horizontal and vertical direction as follows:

$$\boldsymbol{g}_t = \sqrt{\boldsymbol{g}_h^2 + \boldsymbol{g}_v^2} \tag{13}$$

where $\boldsymbol{g}_t$ is the edge map of $\boldsymbol{f}_t$. Due to the pixel value of video sequences rapidly varying over time, each frame will contain different edge complexities. The edge complexity of $\boldsymbol{g}_t$ is measured as follows:

$$\sigma_t = \frac{M \times N}{\sum_{m=1}^{M} \sum_{n=1}^{N} g_t(m, n)} \tag{14}$$

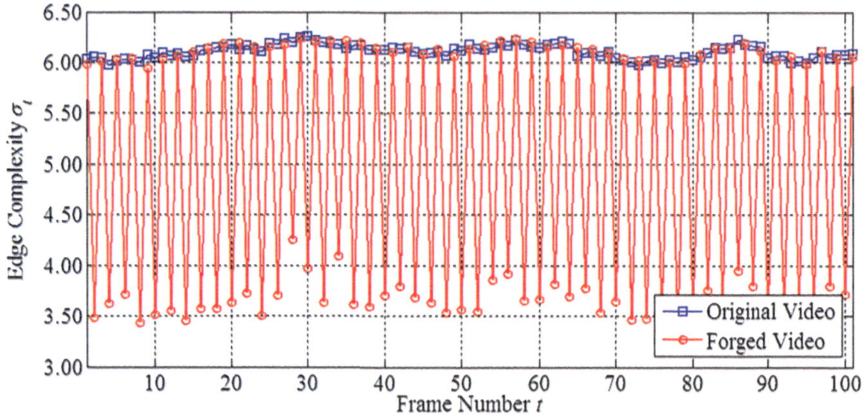

Fig. 5 Edge complexity curves of the *Foreman* video sequence of unforged and forged (Note: The unforged video is the original video of 30fps, and the forged video is transferred to 30fps by the method proposed in [15].)

From Eq. (14), if the edge feature occurs greatly in variation in the intra-frame, the σ_t value is smaller; otherwise it is larger.

Due to MC-FRUC algorithm cannot completely reflect the motion trajectory between adjacent frames, there always are some motion abnormalities in MC-FRUC forgery. Moreover, there are always some artificial traces in the forged frames, which have a large impact on detection edge. Therefore, the edge complexity presents periodic mutations in the forged video. As shown in Fig. 5, for *Foreman* video, using Sobel operator to extract the edge map of unforged and forged video, and then the edge complexity is calculated. It can be seen that the edge complexity curve of unforged video varies steadily and slowly, while the edge complexity curve of forged video rapidly appears in periodic variation. That is why the periodicity of edge complexity curve can be regarded as a strong evidence to detect MC-FRUC forgery.

3.2 Periodicity Detection

By using spectrum analysis to detect the periodicity of edge complexity curve, it can realize the automatic identification of MC-FRUC forgery. Fast Fourier transform (FFT) is used to calculate the spectrum F_k of the edge complexity curve σ_t as follows:

$$F_k = \mathrm{FFT}\left(\sigma_t\right), k = 0, 1, \cdots, L \tag{15}$$

where FFT(•) denotes the FFT operator and L is the length of the noise standard curve. High-pass filter is used to suppress the direct current (DC) and low-frequency

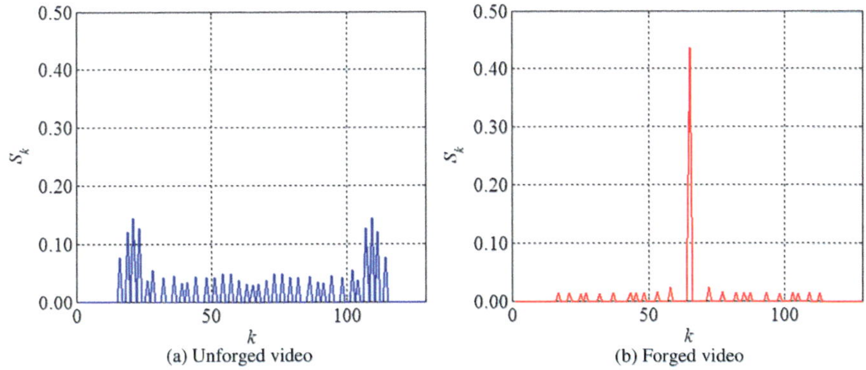

Fig. 6 The spectrum of the edge complexity curve for the unforged and the forged video

(LF) components of spectrum F_k, as follows:

$$F_k^h = \mathrm{HFP}(F_k, d) = \begin{cases} F_k & d < k \leq L - d \\ 0 & 0 \leq k \leq d, L - d < k \leq L \end{cases} \tag{16}$$

where HFP(•) denotes high-pass filter operator, F_k^h denotes high-frequency component, and d is the cutoff frequency. In order to highlight the spectrum peak, amplitude enhancement is performed on F_k^h as follows:

$$S_k = \sum_{l=d+1}^{L-d} \left| F_k^h - F_l^h \right|, k = d+1, \cdots, L-d \tag{17}$$

As shown in Fig. 6, after the processing of Eqs. (16) and (17), the spectrum appears dense and small peaks in the unforged video, while the spectrum center appears a large peak in the forgery video whose amplitude is much higher than the surrounding peak. Thus, the appearance of the large peak proves that the spectrum of edge complexity curve has periodicity. However, it can be seen from Fig. 6b that there are still some small peaks around the large peaks. In order to filter out small peaks, S_k can be disposed as follows:

1. Initialization: after the maximum value of the spectrum, S_k is retained, and the remaining components are set to be 0; ${P_k}^{(0)}$ is assigned, and the iteration variable is set to $n = 1$.
2. Calculate the mean value $E^{(n-1)}$of ${P_k}^{(n-1)}$.
3. Hard threshold shrinkage of ${P_k}^{(n-1)}$ using $E^{(n-1)}$, as follows:

$$P_k^{(n)} = \begin{cases} P_k^{(n-1)} & P_k^{(n-1)} \geq 0.8E^{(n-1)} \\ 0 & P_k^{(n-1)} < 0.8E^{(n-1)} \end{cases} \tag{18}$$

4. If it satisfies

$$\left\{P_k^{(n)}\,|k=0,1,\cdots,L\right\}\cap\left\{P_k^{(n-1)}\,|k=0,1\cdots,L\right\}=\varnothing \tag{19}$$

then stop iteration, output $P_k = P_k{}^{(n)}$; otherwise make $n = n+1$ go to (2) to continue.

As shown in Fig. 7, after the above steps are performed, some small peaks are retained in the unforged video, while only a large peak is retained in the forged video. Therefore, if an abnormal large peak is detected in suspicious video, it can be proved that MC-FRUC forgery operation exists in this video. In order to realize automatic detection, the two spectrum states must be quantified. Thus, the forgery level value is designed as follows:

$$\mathrm{FV}=\frac{\mathrm{MAX}\left\{P_k\,|k=0,1,\cdots,L\right\}}{J\times E^{(0)}} \tag{20}$$

where MAX{•} represents the maximum value of input set, J represents the peak number of P_k, and $E^{(0)}$ represents the mean value of $P_k{}^{(0)}$. As can be seen from Eq. (20), for the forged video, the peak value of P_k is abnormally large and the number of peaks is extremely few, so its FV value is larger. On the contrary, for the unforged video, the peak value of P_k is smaller and the number of peaks is much more, so its FV value is smaller. Thus, the FV value can be regarded as a quantitative indicator to determine whether there is MC-FRUC forgery, and automatic detection can be achieved by setting appropriate threshold value, that is,

$$\text{MC-FRUC}=\begin{cases}\text{on } \mathrm{FV}\geq \mathrm{Thr}\\ \text{off } \mathrm{FV}<\mathrm{Thr}\end{cases} \tag{21}$$

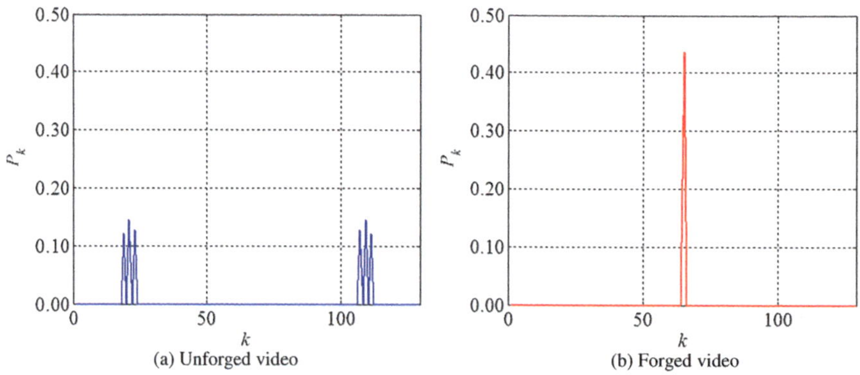

Fig. 7 The spectrum of edge complexity curve after filtering out the small peak interference for the unforged and the forged video

where Thr is the decision threshold, *on* represents the existence of MC-FRUC operation, and off represents the inexistence of MC-FRUC operation.

4 Experimental Results and Analysis

Based on 23 groups of CIF format and 30 fps testing video sequence, the negative set (NS) and positive set (PS) are constructed to evaluate the performance of the proposed algorithm. For the NS, 23 groups of video sequences are directly mixed in Gaussian white noise with standard deviation 0, 3, 5, 7, 9, and 11 to form a total of 138 groups of testing video sequences. For the PS, the video sequence of the NS is first down-converted to 15 fps, and then the MC-FRUC method proposed in [14–16] is adopted to tamper them to 30 fps to form 552 groups of testing video sequences. Firstly, the proposed algorithm is performed to obtain the distribution of FV values of test videos and to determinate the range of FV value of unforged and forged video. Then, the proposed algorithm is used to detect NS and PS, and its performance is evaluated. Finally, we evaluate the ability of the proposed algorithm to resist the attacks of de-noising and compression. The performance index adopts false-positive rate (FPR) and false-negative rate (FNR), respectively, which are defined as follows:

$$\mathrm{FPR} = R_{\mathrm{NS}}/N_{\mathrm{NS}} \tag{22}$$

$$\mathrm{FNR} = R_{\mathrm{PS}}/N_{\mathrm{PS}} \tag{23}$$

where R_{NS} and R_{PS} are the number of outliers in NS and PS, respectively and N_{NS} and N_{PS} are the number of test videos in NS and PS, respectively. Moreover, detection accuracy (DA) is defined as follows:

$$\mathrm{DA} = 1 - \frac{\mathrm{FNR} + \mathrm{FPR}}{2} \tag{24}$$

4.1 Performance Analysis

Figure 8 shows the average FV values of NS and PS under different standard deviation values of noise. As can be seen from Fig. 8, the FV values of NS and PS are significantly different, so the identification of the MC-FRUC can be achieved through hard threshold decision. Figure 9 shows the effects of compression and de-noising attacks on FV values of PS. It can be seen that after the compression and de-noising attacks, the FV value decreases, and it decreases in large amplitude especially for the de-noising attack. The lower the noise variance is, the smaller

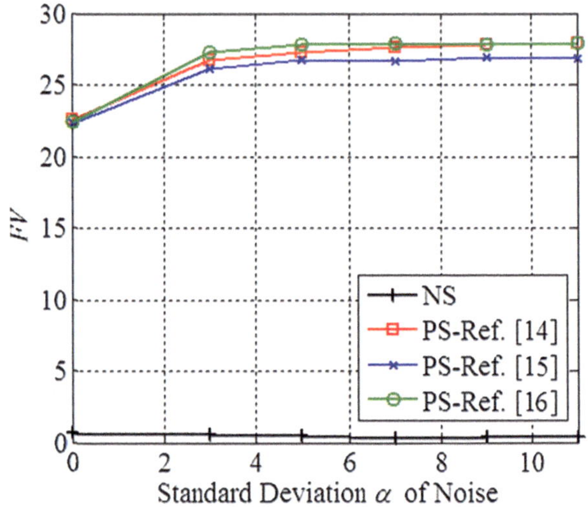

Fig. 8 Under different standard deviations of noise, the average FV value distribution of the NS and PS

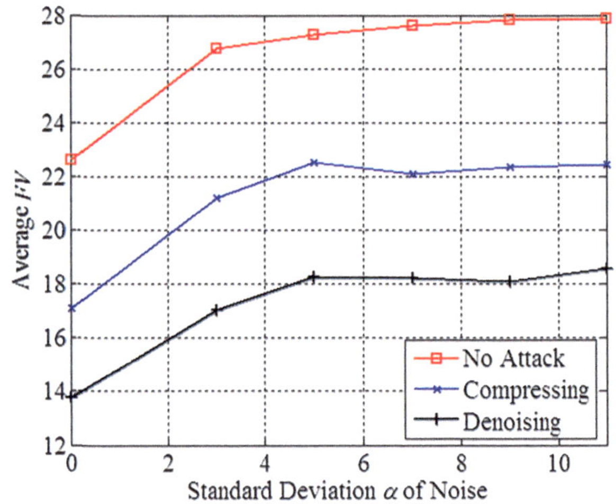

Fig. 9 Effects of compression and de-noising attacks on FV value

the FV value is. While the noise variance is larger, the FV value can still guarantee the higher value. Therefore, the proposed forensics algorithm can better resist the attacks of compression and de-noising.

4.2 Detection Results

The threshold Thr is regarded as a critical decision value of the FV value to determine whether there is MC-FRUC forgery in suspicious video, and it is an important parameter to ensure high DA. In order to set the appropriate threshold Thr, the training video sequence of 23 groups of CIF format and 30 fps different from

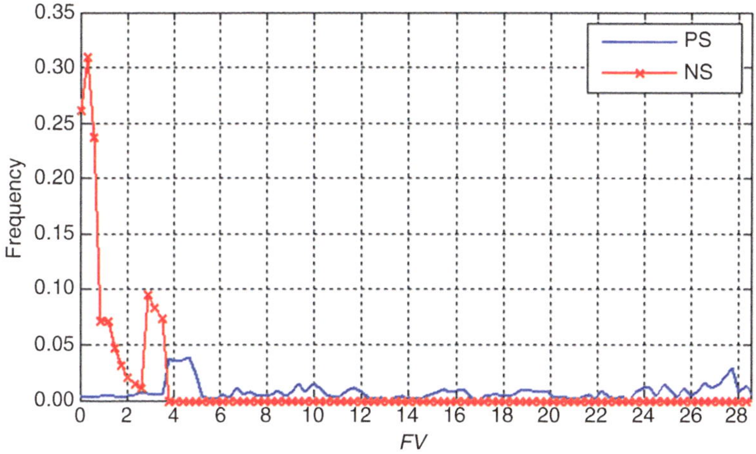

Fig. 10 FV distribution of the NS and PS in training video sequence set

the testing video sequence are selected to form a training set of the same capacity as the test set, in which NS and PS are still constructed by the above method. Figure 10 shows the probability distribution of the FV value of NS and PS in the training set. We can observe that the FV values of mostly NS samples are less than 3. While the distribution of FV values of the PS samples is more uniform and about 90%, the FV values of PS samples are greater than 3. Therefore, a more appropriate threshold Thr should be less than 3. Based on the above analysis results, we select several candidate thresholds in the range of [0.05, 3] with a step size of 0.05 and select the most appropriate threshold among them by cross-validation. First, the NS and PS of the training set are randomly divided into two subsets of the same capacity and nonoverlapping, respectively, one of which is used for training and the other for testing. For the training subset, all candidate thresholds are adopted to detect all samples so as to select threshold of the highest average DA. For the testing subset, the optimal threshold of training subset output is adopted to calculate the average DA of the testing subset. The above cross-validation scheme is executed 10 times, and the variations of threshold and average DA are shown in Fig. 11. It can be observed that the selected optimal threshold can ensure the average DA of test subset above 0.92 in each cross-validation, and the average DA reaches the maximum value when the threshold Thr is 2.4. Based on the cross-validation results, it is more appropriate to set the threshold Thr to be 2.4.

Table 1 shows the average DA of video sequences generated by different MCFI forgery methods. It can be seen from Table 1 that when the suspicious video is not subject to post-processing attack, the average DA reaches 100% under any standard deviation α of noise. This demonstrates that the detection edge complexity is an effective method to identify MC-FRUC forgery. After compression attack is performed on the test video in PS, the detection occur error under the condition of no noise. For example, for the forgery method in [14], the FNR value is 0.16,

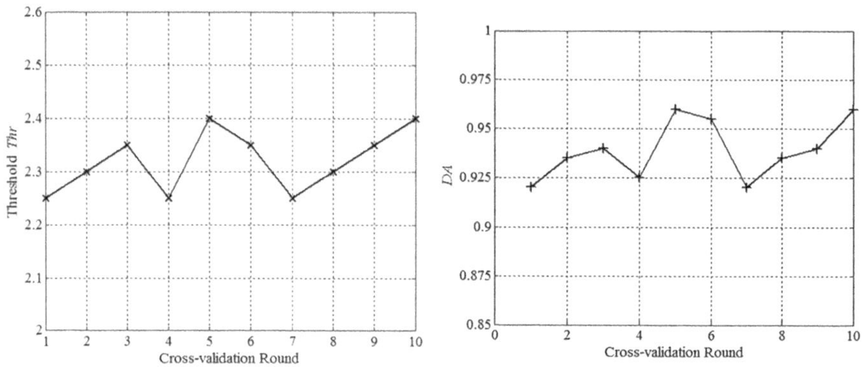

(a) Threshold variation (b) Average DA variation

Fig. 11 Cross-validation results

Table 1 Average detection accuracy of the proposed algorithm for different MCFI forgery methods

MCFI Forgery method	Mixed noise standard deviation α											
	$\alpha = 0$		$\alpha = 3$		$\alpha = 5$		$\alpha = 7$		$\alpha = 9$		$\alpha = 11$	
	FNR	DA	FNR	DA	FNR	DA	FNR	DA	FNR	DA	FNR	DA
	No attack											
Ref. [14]	0.00	1.00	0.00	1.00	0.00	1.00	0.00	1.00	0.00	1.00	0.00	1.00
Ref. [15]	0.00	1.00	0.00	1.00	0.00	1.00	0.00	1.00	0.00	1.00	0.00	1.00
Ref. [16]	0.00	1.00	0.00	1.00	0.00	1.00	0.00	1.00	0.00	1.00	0.00	1.00
	Compression attack											
Ref. [14]	0.16	0.92	0.00	1.00	0.00	1.00	0.00	1.00	0.00	1.00	0.00	1.00
Ref. [15]	0.17	0.915	0.00	1.00	0.00	1.00	0.00	1.00	0.00	1.00	0.00	1.00
Ref. [16]	0.22	0.89	0.00	1.00	0.00	1.00	0.00	1.00	0.00	1.00	0.00	1.00
	De-noising attack											
Ref. [14]	0.17	0.915	0.04	0.98	0.09	0.955	0.13	0.935	0.22	0.89	0.22	0.89
Ref. [15]	0.13	0.935	0.04	0.98	0.09	0.955	0.09	0.955	0.22	0.89	0.13	0.935
Ref. [16]	0.13	0.935	0.04	0.98	0.00	1.00	0.00	1.00	0.00	1.00	0.00	1.00

and the DA is 0.92. However, the average DA recovers to 100% when active noise is implemented, which indicates that the edge complexity can resist the adverse impact of compression attack on the detection for the noisy video. The Gaussian noise is effectively suppressed after the de-noising attack is implemented, which has a certain impact on the proposed algorithm taking the detection edge complexity as the core. In the case of no noise, the FNR value is up to 0.17, and the average DA is only 0.915. The average DA is improved after the noise mixed in, which indicates that the edge complexity can resist de-noising attack to a certain extent for the noisy video.

5 Conclusions

This chapter proposes a forensics algorithm that can identify MC-FRUC forgery by quantifying the variation of edge features in video frames. The proposed algorithm first using the Sobel operator to detect edges of video frame and then the edges is quantified to obtain the edge complexity of each frame. Finally, the periodicity of the edge complexity along time axis is detected, and MC-FRUC forgery is automatically identified by hard threshold decision. Experimental results show that the FV values of unforged videos are much lower than those of forged videos. When the decision threshold is set to be 2.4, the proposed algorithm shows a good detection performance on the NS composed of 138 video sequences and the PS composed of 532 video sequences. When the post-processing attack is not performed on testing videos, the average DA reaches 100%. The proposed algorithm can still maintain the stability of the detection accuracy after the attacks of de-noising and compression.

References

1. Choudhury, S., Al-Turjman, F., & Pino, T. (2018). Dominating set algorithms for wireless sensor networks survivability. *IEEE Access, 6*(99), 17527–17532.
2. Al-Turjman, F., & Alturjman, S. (2018). 5G/IoT enabled UAVs for multimedia delivery in industry-oriented applications. *Springer's Multimedia Tools and Applications, 1*, 1–22.
3. Al-Turjman, F. (2018). QoS–aware data delivery framework for safety-inspired multimedia in integrated vehicular-IoT. *Elsevier Computer Communications, 121*, 33–43.
4. Tsai, T. H., Shi, A. T., & Huang, K. T. (2016). Accurate frame rate up-conversion for advanced visual quality. *IEEE Transactions on Broadcasting, 62*(2), 426–435.
5. Bian, S., Luo, W., & Huang, J. (2014). Exposing fake bit rate videos and estimating original bit rates. *IEEE Transactions on Circuits & Systems for Video Technology, 24*(12), 2144–2154.
6. Bian, S., Luo, W., & Huang, J. (2014). Detecting video frame-rate up-conversion based on periodic properties of inter-frame similarity. *Multimedia Tools and Applications, 72*(1), 437–451.
7. Wang, Z., Bovik, A. C., & Sheikh, H. R. (2004). Image quality assessment: From error visibility to structural similarity. *IEEE Transactions on Image Processing, 13*(4), 600–612.
8. Yang, J., Huang, T., & Su, L. (2016). Using similarity analysis to detect frame duplication forgery in videos. *Multimedia Tools and Applications, 75*(4), 1–19.
9. Choi, D., Song, W., & Choi, H. (2015). MAP-based motion refinement algorithm for block-based motion-compensated frame interpolation. *IEEE Transactions on Circuits & Systems for Video Technology, 26*(10), 1789–1804.
10. Bestagini, P., Battalia, S., Milani, S., Tagliasacchi, M., & Tubaro, S. (2013). Detection of temporal interpolation in video sequences. In: *IEEE International Conference on Acoustics, Speech and Signal Processing*, pp. 3033–3037.
11. Yao, Y., Yang, G., & Sun, X. (2016). Detecting video frame-rate up-conversion based on periodic properties of edge-intensity. *Journal of Information Security & Applications, 26*, 39–50.
12. Xia, M., Yang, G., & Li, L. (2017). Detecting video frame rate up-conversion based on frame-level analysis of average texture variation. *Multimedia Tools & Applications, 76*(6), 8399–8421.

13. Ding, X., Yang, G., & Li, R. (2018). Identification of motion-compensated frame rate up-conversion based on residual signal. *IEEE Transactions on Circuits & Systems for Video Technology, 28*(7), 1497–1512.
14. De, H. G., Biezen, P. W. A. C., & Huijgen, H. (1993). True-motion estimation with 3-D recursive search block matching. *IEEE Transactions on Circuits & Systems for Video Technology, 3*(5), 368–379.
15. Yoo, D. G., Kang, S. J., & Kim, Y. H. (2013). Direction-select motion estimation for motion-compensated frame rate up-conversion. *Journal of Display Technology, 9*(10), 840–850.
16. Liu, H., Xiong, R., & Zhao, D. (2012). Multiple hypotheses Bayesian frame rate up-conversion by adaptive fusion of motion-compensated interpolations. *IEEE Transactions on Circuits & Systems for Video Technology, 22*(8), 1188–1198.
17. Jeong, S. G., Lee, C., & Kim, C. S. (2013). Motion-compensated frame interpolation based on multi-hypothesis motion estimation and texture optimization. *IEEE Transactions on Image Processing, 22*(11), 4497–4509.
18. Kanopoulos, N., Vasanthavada, N., & Baker, R. L. (2002). Design of an image edge detection filter using the Sobel operator. *IEEE Journal of Solid-State Circuits, 23*(2), 358–367.

A Neuro-fuzzy-Based Multi-criteria Risk Evaluation Approach: A Case Study of Underground Mining

M. F. Ak

1 Introduction

The Internet of Things (IoT) is defined as a unique network around the world, where uniquely addressable objects are formed and the objects in that network communicate with each other through a specific protocol. It is also possible to define this concept as a system of devices that communicate and communicate with each other through a variety of communication protocols and by sharing information with an intelligent network [1]. Since safety is one of the most important and prone concepts for occupational safety and health in mining, proactive based IoT applications supported risk analysis can be adapted to the system in mining sector [2]. IoT sensors and devices can be used to improve total safety level [3]. Investigation of proper way with wireless sensor network features is done to monitor outdoor environment [4]. According to wireless network ability, changing environment conditions can be explored and adopted [5]. The system can help to minimize hazards and environmental problems such as water pollution and fires. Environmental monitoring with IoT can also provide cost-efficient and proper solutions [6].

Risk analysis of any workplace is first step to check and observe total safety level. There are many quantitative and qualitative risk assessment tools to analyze reasons of accidents in various industries and features of the workplace conditions in sources. Multi-criteria decision-making-based method is one of the most important methods [4]. In these methods, decision-makers often have difficulty in accurate rating and assessment in a danger related with risk parameter. Therefore, implementing potential risk assessment methods can show satisfactory results in terms

M. F. Ak (✉)
Industrial Engineering, Antalya Bilim University, Döşemealtı/Antalya, Turkey
e-mail: fatih.ak@antalya.edu.tr

F. Al-Turjman (ed.), *Artificial Intelligence in IoT*,
Transactions on Computational Science and Computational Intelligence,
https://doi.org/10.1007/978-3-030-04110-6_9

of incomplete risk data or high uncertainty. In this study the ways that combine multi-criteria decision-making method and fuzzy sets are accepted to model the structure [5]. One of the important advantages of fuzzy multi-criteria decision-making methods is relatively assessing the risk parameters by using fuzzy numbers instead of insignificant numbers [6]. Chart 1 shows a comparative summary about a range of new researches related with fuzzy multi-criteria decision-making methods in occupational health and safety risk assessment [7–15].

In the underground mining, the work environment becomes more appropriate with the OHS, and the occupational accidents and occupational diseases are affected positively by taking necessary precautions, by the worker's feelings of trust originating from the work environment. In addition, these factors become the reasons for the improvements that cause positive effects as in the employee's daily life as his efficiency in workplace [16]. This is also important in the aspect of improvement and continuity of the system in a similar way.

From this point of view, potential risks in mines within occupational health and safety and forming OHS risk analysis scheme and managing the necessary precautions actively by classifying emerging dangers are aimed in this study in the process of carrying out the underground mining. To improve the system precautions which are taken necessarily in accordance with detailed risk analysis.

As a part of this study, the risk assessment study based on fuzzy and neuro-fuzzy logic is carried out by using the data based on OHS risk analysis of a metal mine industry which has more than one billion production capacity and is compared to failure level of the basic risk assessment methods.

1.1 Original Contribution

Based on the data that is obtained within the study, by using neuro-fuzzy logic approach in underground mining, risk analysis is done and risk levels are estimated. Underground mining risk parameters in OHS assessing by neuro-fuzzy logic are used for the first time, and the results are compared with the results of basic risk assessment methods in the point-of-failure criterion.

Besides that, L-matrix risk assessment results and two-phase risk analysis results [17] obtained by Buckley fuzzy analytic hierarchy method and fuzzy VlseKriterijumska Optimizacija I Kompromisno Resenje (VIKOR) methods are compared with neuro-fuzzy logic obtained by the results from risk assessments by considering the failure measures in OHS; in terms of implementing more detailed studies, the sources in this area have been enriched.

Risk assessment studies based on two or three parameters such as severity, possibility, and noticeably contributed to the resources in terms of dealing with the risks more effectively by adding different parameters that extend the content like precision of using personal protection tools, precision for not to renew the tools and ignoring the results.

The World Health Organization defines the OHS term as multidisciplined operation including four basic topics: (1) to prevent occupational accidents by

Chart 1 Fuzzy multi-criteria decision-making methods and OHS risk assessment studies and comparison

Study	Aims	Method	Approaches
Grassi [16]	Ranging of a famous Italian sausage brand's risky actions in production process	FTOPSIS	Ordering of hazards with FTOPSIS method by considering the effect of human behavior and environment on risk level instead of ordinary extend probability factor
Gul and Guneri [18]	Prioritizing the risk groups in aluminum plate factory	Buckley's FAHP. FTOPSIS	FTOPSIS and FAHP are used for ordering the measure of two criteria which is obtained from decision matrix method and the risk types of each department in factory and identifying its method
Mahdevari [29]	Searching the risks related with health and safety in underground coal mines	FTOPSIS	FTOPSIS method is used for organizing mine risks in Iran
Liu and Tsai [28]	Providing the risk analysis rates of risk factors in construction sector	QFD, FANP, fuzzy failure mode and effect analysis (Fuzzy FMEA)	QFD is used to show the relation between construction items, risk types, and risk factors. FANP is used to identify risk types and risk factors. Fuzzy FMEA is used for risk factors that are based on fuzzy inference approach analyzing risk rate
John [23]	Suggesting new fuzzy risk assessment approach to ease of understanding the obscurity in harbor operations	FAHP, ER	FAHP is used for analyzing the complex structure of harbor operations, and identifying the severity of the risk factors and ER is used for synthesizing these results
Gul and Ak [17]	Propose an outline for OHS risk assessment in mining industry with comparison	PFAHP, FTOPSIS, circumcenter of centroids	Used PFAHP to weight risk parameters and FTOPSIS and circumcenter of centroids to rank hazards
Hu [21]	Risk analysis of green constituents for hazardous substance	FAHP, FMEA	FAHP is used to identify relative rate of the four factors. FMEA is used to calculate the risk priority number
Ebrahimnejad [12]	Risk identification and assessment for build-operate-transfer projects	FTOPSIS, FLINMAP	FTOPSIS and FLINMAP are used to order the high risks in build-operate-transfer projects
Djapan [10]	Identification of risk levels in small- and medium-scale manufacturing businesses	FAHP	FAHP is used to define hierarchical structure, sub-factors, and the importance of relative rate of the sub-factors
Vahdani [33]	Searching the reasons for failure of the steel production project	FMEA, TOPSIS	Parameters of the FMEA are used to identify fuzzy logic TOPSIS and the reasons of failures

FTOPSIS Fuzzy technique for order preference by similarity to ideal solution, *PFAHP* Pythagorean fuzzy analytic hierarchy process, *AHP* Analytic hierarchy process, *ANP* Analytic network process, *QFD* Quality function deployment, *TOPSIS* Technique for order preference by similarity to ideal solution, *PROMETHEE* Preference ranking organization method for enrichment of evaluations, *FAHP* Fuzzy analytic hierarchy process protecting and improving employee's health; (2) improving and supporting healthy and safe work, workplace, and work organizations; (3) increasing employees' physical, mental, and social welfare; and (4) enabling the workers having fruitful lives both socially and financially [18, 19].

1.2 Occupational Health and Safety

OHS is all the studies and analyses that aim reducing the problems which may occur in all processes related with work and employees. In terms of OHS, structural difficulty of the underground mining services and precision as of divergent and rapidly changing equipment structure need to be done detailed research and analysis [20]. In workplaces employees work that are of the most important parts of underground mining services, being with unreliable factors in the aspect of OHS and lack of precautions for these factors are required to create different regulations. The underground mining organizations are part of the most dangerous group in terms of occupational accidents and occupational diseases in OHS risk classification list, and this makes compulsory to conduct detailed and effective risk analysis of underground mining services.

Making workplaces appropriate for OHS in underground mining, identifying the occupational accidents and occupational diseases, taking necessary precautions, and making employees feel safe due to working environment affect their works positively [21]. In addition these factors become the reasons for the improvements that cause positive effects as in the employee's daily life as his efficiency in workplace. This is also important in the aspect of improvement and continuity of the system in a similar way.

Based on these results, potential risks in mine organizations within occupational health and safety and forming detailed OHS risk analysis scheme and managing the necessary precautions actively are aimed in this study in the process of carrying out the underground mining. To improve the system precautions which are taken necessarily in accordance with risk analysis, results are stated.

1.3 Occupational Health and Relevant Terms

There are many definitions of the health term in different sources. WHO which does practices on health internationally defines the health as "not only health" or disablement situation, but also, it's a person's wellbeing and feeling completely in safe physically, mentally, and socially [22]. This definition may change depending on country, culture, and customs. Criteria of a person being healthy cannot be explained by having physical disability or catching a disease. Besides physical aspects, being safe mentally and socially are the other criteria that make a person healthy. From this point of view, health term is the basis of occupational health

[23]. During the employee's process of being at work or workplace, not being exposed to the risks that will threaten his health and being protected from this are the occupational health term's basic relation with health. Protecting from the dangers due to the employees' working environment and job structure is the essential point in the definition of the occupational health term.

Occupational health is a term that searches the physical and mental situation of the employees in workplaces and aims to improve the conditions. In the aspect of occupational health, it is the most important term not to risk and protect the right to life which is among an individual's natural and fundamental rights, causing from the workplace [24]. Besides that, it also is a compulsory part of the occupational health system to have high efficiency, increase quality standards, and create a sustainable working environment. Comprehending this term as a social culture rather than understanding it as a term just related with workplaces is significant not only in the aspect of employer-employee interaction, but also it is important for developing healthy and safe environment by all society ranks [25]. There are many national and international scale regulations with the aim of increasing the OHS culture. An individual's work department is also substantial in accordance with explanations of the occupational health term. In terms of OHS, while doing the work that employee is in charge of, he needs to be conscious and knowledgeable, and to be able to do that, person's nature and kind of the job needs to adjust in maximum possibility [26]. It is necessary that employees work in appropriate jobs which are suitable with their education level and physical and mental state with regard to actualizing the aims of the occupational health term. If a person works in a suitable job in that sense, opportunities of protecting from the risks will increase, and the number of the occupational accidents and its effects will be able to decrease in parallel with this.

1.4 Occupational Safety and Relevant Terms

Safety term expresses being secure. Occupational safety states that taking precautions against the risks that employees can face. Among the process of employees being at work or workplace, the unrepairable consequences that may occur because of occupational accidents show the importance of creating a safety culture term [27]. This must be changeable according to the business and job structure, but it should be aimed at spreading it to a whole society, as well. For instance, it must be aimed that while dealing with workers who are having health problems, whether work process is safe or not for underground mining sector which is considered as an important situation in this study. A worker who is not feeling safe in the working environment makes mistakes, and troubles occur in the job unavoidably [28]. Safety culture should be provided with employee-employer cooperation, and drafts, plans, and programs that serve this should be prepared periodically. It is not possible that being successful in a safe environment is constituted by only an employee or employer. Systematically, all the pieces should have effects on it and contribute to form the safety culture [29, 30].

Many occupational accidents occur every year all around the world due to the deficiencies in the implementation of OHS. According to the International Labor Organization data, more than one billion people die every year because of occupational accidents, and more people are physically or mentally hurt due to these accidents [31]. These data show how the OHS issue is significant. Based on the statistical data which is brought out by social security institution, recently, the people who die because of occupational accidents and occupational diseases have been increasing, and for this reason, it has been observed that the people who become permanently disabled are increasing, too [32]. Fatal occupational accidents increased more than 10%, when it is compared to the previous year, and in parallel with this, being permanently disabled and occupational diseases highly increased in 2014. In that case, according to statistical studies done for underground mining, currently, Turkey is rather behind the world in terms of the number of the occupational accidents in developed countries, number of the people who is permanently disabled, and number of the occupational disease rates. By taking all these into consideration, legislature regulated the laws and passed the 6331 numbered law under the title of OHS. Before passing a law special for OHS, there were legal analyses related with occupational accidents and occupational diseases by different regulations. For example, there were analyses under separate and different laws for the pecuniary and nonpecuniary damages that occur after an occupational accident. This can be resulted with faults, and it may not be possible to get positive results. Besides that, a legal obligation that includes OHS regulations and adapts it emerged in point of the European Union (EU) adaptation process and ILO directives. OHS term is suitable for chances and needy for improvement according to gathered statistical occupational accident and occupational disease data in Turkey in recent years. OHS policies are changed, and current laws are shaped in accordance with EU and ILO directives [33]. Before the legalization, OHS terms that are emphasized on the specific parts of the different laws are combined in one title.

Following the explanations of the health and safety concepts that generate the OHS concept, the differences between these two concepts have been revealed. Occupational health refers to search for developing and improving their physical and mental status, as occupational safety at work or in the workplace is the protection of employees from risks and the risks they face in the workplace. The fact that employees are in a healthy and safe environment is the main objective of these two concepts [34]. A healthy environment in terms of OHS increases the productivity of employees and the ability to work in a safe environment. Proper physical, psychological, and social workplace environment to be provided for employees is a proactive structure and is the priority of identifying and minimizing risks in advance. It should not be forgotten that the budget for the precautions to be taken by proactive approach to work accidents is much more appropriate and necessary than the damages that will happen as the result of occupational accident.

1.5 Management Concept in OHS

It is compulsory to have a managerial understanding in the system in terms of OHS structure. It would be normal to occur and growing problems in terms of occupational safety and occupational health in a business that the management is not being part of and support OHS implementation and activities. Studies on OHS indicate that the lack of a management-supported system is one of the most important reasons for the increase in job accidents. This demonstrates the importance of management understanding in the success of OHS practices. The implementation of a well-designed and experienced OHS management model means that employees are aware of the hazards and the risks are minimized; all employees are in a safe and healthy environment as a team. In addition, the productivity of employees and work will also improve. In this context, the employer must make investments for the occupational area of the OHS without any legal obligations on occupational health and safety and should not regard this as an unnecessary cost. Protecting from danger is much less costly than protection from possible consequences. It should also be kept in mind that when an appropriate environment for OHS is created, it is possible to achieve a sustainable success by reducing the number of mistakes relatively.

In places where work will be done, the target risks should be determined in advance in terms of OHS management, and occupational accidents and occupational diseases should be prevented [35]. These undesirable consequences may cause workforce and financial loss in the workplace. This workplace-based problem may cause social trauma without staying local. Soma disaster that happened on 13 May 2014 is an example of this. The human loss was far beyond financial loss, and it is clear that it is not an easy situation to overcome this trauma. As it is frequently written at workplaces and working points, a safe working environment must be created in actual implementation and management decisions. Management should inform and provide periodical trainings on OHS issues to employees. In addition, each division-specific OHS chargehand should be identified and known by all employees. An OHS plan which is prepared by the management should also be periodically checked and updated. A safe and healthy workplace means healthy and efficient work and ultimately healthy functioning. Each part of the OHS management practices should include risk assessment and practices, and teamwork should be adopted.

2 Risk Assessment in OHS

Concepts of safety, health, and welfare of whole industries' employees in the workplace are covered by OHS [36]. Due to the poor management of OHS, the underground mining sector that is one of the most affected industries aims to raise the health and safety standards in the relevant institutions in Turkey and all around the world. Many unique hazards can be observed in mines that can potentially affect the health of workers in their departments [37]. These risks increase the accidents in the workplace, affect both the safety of employees and the environment antagonistically, and reduce the efficiency and work performance. OHS practices are not fully implemented in practice, and necessary diligence has not been shown for the underground mining industry in Turkey. There is a negative correlation in economic success and occupational success in Turkey although there should be a positive correlation between OHS practices and economic success. In order to be able to see the OHS policy and compliance in the mining sector more clearly, the new OHS 6331 numbered law also stipulates the necessity of risk assessment. As a result of this reason, this study aims to provide a neuro-fuzzy logic-based OHS risk evaluation for underground mining personnel and to contribute to the development and planning of underground mining services.

On the other hand, OHS risk assessment has several approaches in the planning and management of the underground mining industry. In a literature study, the VIKOR method has been proposed to prioritize failure modes for a fuzzy FMEA based on fuzzy set theory and for general anesthesia process risk assessment [38]. One of the most common risk analysis methods is FMEA which has three parameters. In this study parameters of the FMEA are weighted by fuzzy set theory, and the risk priorities of error modes are determined by the extended VIKOR method. In another study [39] a new FMEA approach was implemented with the fuzzy analytic hierarchy process (FAHP), entropy, and fuzzy VIKOR methods in the production process of a mine. In one other study [40], a three-stage risk-based prioritization scheme is suggested to select the best care strategy for prioritizing medical devices. First, the fuzzy FMEA was implemented by taking into account various risk assessment factors. Second, seven dimensions are used to consider every aspect of hazards and risks, such as use-related hazards, age, and usage. Finally, a simple method has been used to determine the best maintenance strategy for each device based on the scores of the previous two steps. A new risk priority model has been proposed to assess the risk of failure modes based on the fuzzy set theory and the MULTIMOORA method [41]. Besides, a case study on preventing baby abduction was also presented.

Based on the studies in the literature, it is understood that the current study has contributed to the relevant resources on some aspects of underground mining risk assessment.

A fuzzy logic-based structure that removes the shortcomings of a clear risk parameter calculation and reduces the inconvenient has been used to make decisions. Unlike classical OHS risk assessment methods, decision-makers choose parameter weights with a neuro-fuzzy tongue scale. Decision makers can get benefit from neuro-fuzzy logic while establishing appropriate structure of risk assessment process neuro. Classical OHS risk assessment methods consist of either two risk parameters (e.g., decision matrix method) or three risk parameters (e.g., Fine-Kinney method, FMEA method). Increasing the number of parameters in the study will provide a more flexible usage, and a consistent and broad risk assessment model will be presented. This study is the first risk assessment study in the country where the neuro-fuzzy logic approach in occupational health and safety is applied in underground mining.

2.1 OHS Studies in Underground Mining and Current Status

Developing the standards of health and safety at the associations related to underground mining is aimed. Mines which are one of the largest employment groups in the manufacturing sector are facing serious hazards due to the risks which are technically defined previously. These risk factors are increasing the number of occupational accidents, affecting the safety of both the employees and the nature, decreasing the productivity and work performance. Although Turkey has been making significant progress in economic field and making relatively rapid growth in its own geography, OHS implementations are not carried out in a complete and active way, and adequate care has not been shown in underground mining industry. Hence, the aim of the study is to assess the risks for underground mining staff, contribute to process planning in mining, and develop the regulations. There should be an OHS policy which is approved and carried out by the executives of all the enterprises, planned periodically, and open to development and updates [42].

Planning is the first stage of OHS policy. Within the frame of this plan, all the hazards should be defined in terms of facility structure and workplaces, and the staff should be informed about this. Responsible staff should be designated for special OHS security in each department; permanent and sustainable success should be aimed. In addition to this, convenient regulations should be prepared in accordance with the facility's structure in the medium-sized and large-scale enterprises, and template plans should be created [43].

In order to make staff learn about these regulations, trainings and seminars should be organized and tested periodically. By OHS policy, control mechanism matters. As a result of the control understanding which has gained more importance in underground mining, it has been playing a significant role in precluding the occupational accidents which could result in death. When researching the Ministry of Labor and Social Security (MOLSS)'s short-term and medium-term policies, it is stated that the main aim is to establish a data archive regarding OHS and assess these data on periodical basis [44]. As a result of these studies which will be done on sectoral basis, there will arise a better opportunity to fight against the occupational accidents and illnesses.

As exemplified in the second part, Turkey needs more active policies and follow-ups in order to move to higher positions from where it is currently in OHS field. In the studies done, it is seen that there are risks in underground field, inadequacy, and malfunctions in diagnosing occupational illnesses. The top reason underlying this inadequacy is that OHS training is not in the required position that it should be. In the facilities which carry top level of risk in hazard classifications such as underground mining, the number of training sessions should be increased and conceptualized as project. In underground mining risk assessment study, it was found that the participation and interest of the employees in trainings and researches were low. It is obvious that Turkey needs achievements in OHS policies to reach at the position aimed within the scope of Vision 2023. At this point, it would be rightful to invest in R & D works on both state and private sector basis. In order to form OHS policy as a culture, both enterprises and employees are needed to be encouraged to participate in the cooperation.

Like it happens in all the businesses, the management system is the key to reach the success in a workplace. At the same point, OHS management implementations are required to be prepared in an organizational way and be efficient in all businesses. In order to carry out the responsibilities of OHS in an efficient and successful way, management support is mandatory. These are also valid and critical for the underground mining-based OHS management system [20]. Feeling safe and being healthy for the staff working in underground mining services will lead to receive better results in studies. As the nature of the job, mining workers face many hazardous situations every day, and they try to avoid these situations as much as possible. Individually, it is not possible to keep away from the hazards in underground mining sector. Hence, establishing OHS management and control system efficiently will lead to success at the workplace and get protected against the hazards in a more efficient way.

2.2 *Risk Assessment in Multi-criteria Decision-Making Methods*

Risk assessment involves the data of the risk-generating elements being finalized by calculating and the analysis of their possibilities. The risk analysis, which is basically divided into two as qualitatively and quantitatively, can be actualized more efficiently by expressing more criteria than one of the complex systems thanks to the multi-criteria decision-making methods. These methods enable the decision-makers to get decision results as easily, quickly, and efficiently as possible, in case that there are multiple alternatives and multiple parameters.

2.2.1 5×5 Matrix Risk Assessment Methodology

Risk assessment is an approach, widely used in OHS risk analysis, which measures the 5×5 matrix method risk classification, also known as the decision matrix technique, based on the decision that reviews it within the dimension of both severity and probability parameters. With this method, by multiplying severity and probability, a risk value criterion is obtained.

Initially, the measurement of severity and probability is determined by this method (Charts 2 and 3). Afterward, risk matrix and decision-making table are

Chart 2 Probability gradation chart [18]

Probability (probability value)	Gradation steps for likelihood of appearance
Very low (1)	Almost never
Low (2)	Very rare (once a year), only under abnormal conditions
Medium (3)	Rare (a few times a year)
High (4)	Often (once a month)
Very high (5)	Very often (once a week, every day), under normal working conditions

Chart 3 Severity gradation chart [18]

Severity (severity value)	Gradation
Very mild (1)	No loss of working hours, can be eliminated, requires first aid
Mild (2)	No loss of working day, ambulatory treatment without lasting consequences
Moderate (3)	Mild bodily harm, inpatient treatment/injury
Severe (4)	Severe injury, long-term treatment, occupational illness
Very severe (5)	Death, permanent total disablement

Chart 4 Risk score evaluation matrix [18]

Risk score	Severity				
Probability	1 (Very mild)	2 (Mild)	3 (Moderate)	4 (Severe)	5 (Very severe)
1 (Very Low)	Insignificant 1	Low 2	Low 3	Low 4	Low 5
2 (Low)	Low 2	Low 4	Low 6	Medium 8	Medium 10
3 (Medium)	Low 3	Low 6	Medium 9	Medium 12	High 15
4 (High)	Low 4	Medium 8	Medium 12	High 16	High 20
5 (Very high)	Low 5	Medium 10	High 15	High 20	Intolerable

created (Charts 4 and 5). The acceptability of the risks is interpreted according to Chart 5.

In the study, a fuzzy approach was proposed that allows experts to evaluate two parameters of the 5×5 matrix method using linguistic variables:

(1) To overcome deficiencies in calculating a clear risk score and (2) to reduce inconsistencies in decision-making. Since the conventional 5×5 matrix method is based on the weight of an equal measure for the severity and probability, it can cause limitations and inconsistencies.

3 Fuzzy Logic and ANFIS

In 1940s, neurophysiology expert Warren McCulloch and mathematician Walter Pitts conducted a research on how the neuros might work. This study is accepted as a first study on neural networks. Their study resulted with a simple neural network

Chart 5 Risk score assessment matrix [18]

Intolerable unendurable risks [25]	Work shouldn't be started until the identified risk can be lowered to an acceptable level; if there is a continuing action, it should be stopped immediately. If it is not possible to lower the risks even though precautions are taken, then the action should be prevented
Important risks [15, 16, 20]	Work shouldn't be started until the identified risk can be reduced; if there is a continuing action, it should be stopped immediately. If the risk is about the continuation of work, then urgent measures should be taken, and as a result of these measures, the continuation of the activity should be decided
Medium risks [8–10, 12]	Activities should be initiated to reduce the identified risks. Risk mitigation measures can take time
Tolerable risks [2–6]	Additional control processes might not be needed to reduce the identified risks. However, existing controls should be continued, and they should be supervised
Unimportant risks [1]	It might not be necessary to plan control processes and keep the records of the actions to be done, to dispose the identified risks

by using electric circuits, to describe how the neurons in the brain work. In 1950s, with the advances in computers, it became possible to simulate the hypothetical neural networks. However first studies resulted unsuccessfully. In the 1960s, the models named “ADALINE” and “MADALINE” were developed. ADALINE was developed with the aim of recognizing binary patterns, while MADALINE was applied firstly on neural network to a real-world problem. It is used firstly on an adaptive filter which supplies opportunity to eliminate echoes on phone lines. The model is still in use commercially. After 1970, development of artificial neural network (ANN) accelerated and brought solutions to important problems. The development of correlation matrix memory and multilayer perceptron were some of the developments after 1970. Today, neural networks are used in many areas. The fundamental idea of the neural networks is that, if it works in nature, it should work on computers as well. ANN has a strong nonlinear mapping ability, strong learning ability, and a high sensitivity level. An artificial neural network is a classifier modeled after the human brain. The human brain has billions of nerve cells (approximately ten billion). They form a complex and developed signal transmission network system by connecting to homogeneous cells. Cell gathers the input from all the connection points until it reaches to a certain level. After the input reaches this level, the cell sends a signal to all the cells it is connected to in the network. ANN is expected to imitate the human brain, and its working is the same as the biological neuron structure in the human brain. However, the working of ANN is based on mathematical evidence. Some neural network structures are not close to the brain in location, and some doesn't have a biological counterpart in the brain. At the same time, neural networks have a very strong resemblance to the biological brain, and therefore neuroscience terminology is used frequently. ANN learns the relationship between the selected input and the output. Neuros are connected to each other to help this process. Each neuron processes the input taken from one

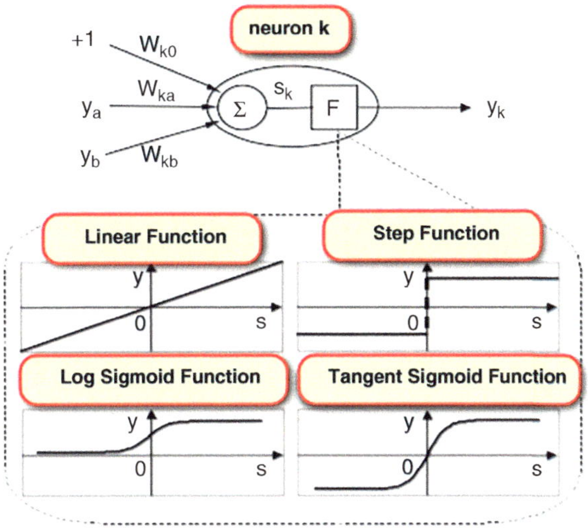

Fig. 1 Basis of an artificial neuron

or more neurons and produces output. The ANN structure is considered to be a versatile framework in which learning progresses by taking into account external or internal data passing through the system. ANN can take a few weighted inputs and summarize them, and if the combined inputs exceed a threshold, they will activate and send an output. Activation process determines which output to send, and the output is usually between 0 and 1 (or −1 and 1). Since the derivatives of the activation function are often used in network training, a few additional calculations are needed so that the derivative can be expressed by the original function value.

Although the artificial neural network is simulated from the biological model, its base unit has a simpler structure. Figure 1 shows the basis of an artificial neuron.

y_a and y_b input w_{ks} represent the connection weight of each input. w_0 is the transfer function with a value of +1 to every neuron. This function is used as a transfer function between the input layer and the hidden layer in the system. It regulates the system output with interval [−1 1]. In the next step, mathematical calculations are conducted, majority of addition and the products of process are fed through nonlinear transmission functions to produce output. The structure of all the artificial neural network is the same, but they may differ at some points.

The most common transfer functions were shown in Fig. 1. Depending on the application, transfer functions may vary. The linear function can be selected to solve linear problems, and the result from the summation function is multiplied by a certain coefficient and calculated as output. This can be expressed as y = c.s, where the c value is invariant. Threshold function gets the value of 1 or −1. At this point, value is determined according to the set threshold level.

This can be shown as:

$$y = \{1 \text{ if } s \geq 0, y = -1 \text{ if } s < 0\} \tag{1}$$

The sigmoid activation function is a continuous function that can be derived. While using sigmoid activation function, artificial neuron may seem like a natural neuron. This function produces output between 0 and 1.

$$y = 1/\left(1 + e^{-s}\right) \tag{2}$$

Tangent sigmoid function is a function which is similar to sigmoid function. The difference is the point at which the output values change [−1 to 1] by giving a negative output.

$$y = \left(e^{s} - e^{-s}\right) / \left(e^{s} + e^{-s}\right) \tag{3}$$

While artificial neural nets are but artificial interconnections of neurons, they can learn the relationships between the inputs and outputs selected from previous experience. ANNs can also fulfill their task as parallel processing. This gives an advantage in terms of speed. A typical ANN can describe and learn the relationships between the input and the output of a nonlinear multidimensional system. ANN and the structure of multidimensional nonlinear system are shown in Figs. 2 and 3.

The main problem with the single-layered network is that they cannot classify the nonlinear functions. The complexity of the single-layered neurons isn't large enough to address bigger classification problems. The logical depth of such problems is too great for an only perceptron to overcome. To overcome this problem, a few layered networks should be formed. These are called multilayer perceptron. The structure of the multilayer perceptron is shown in Fig. 4.

x_i represents the input, w_{ij} represents the weight, y_k represents the output, and F represents the transfer function.

ANNs are superior to the conventional algorithms in terms of processing style and speed, storage structure and network prevalence, learning ability, offering error tolerance level, making use of experience, and flexibility of rule-setting.

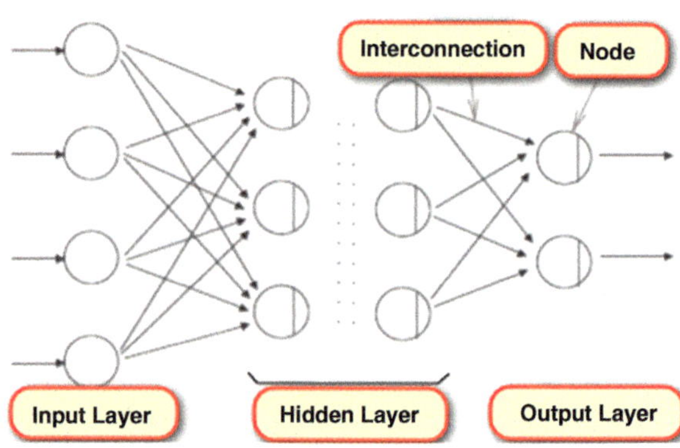

Fig. 2 Artificial neural network

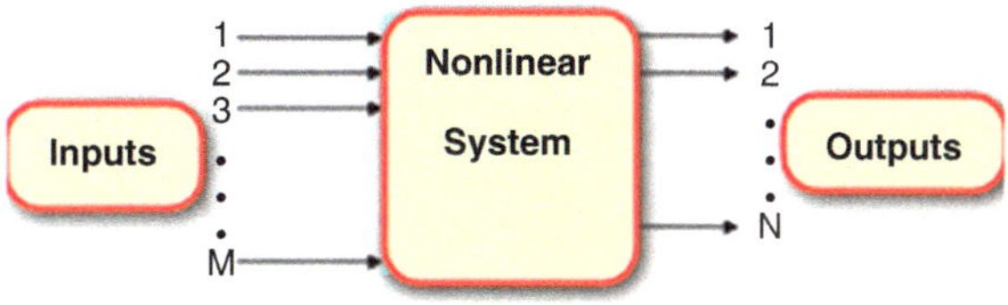

Fig. 3 Nonlinear multilayered structure

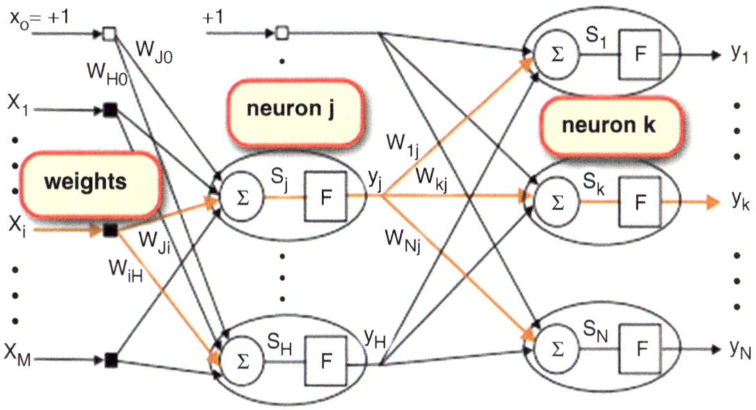

Fig. 4 Multilayer perceptron structure

ANN is widely used in fields such as space industry, electric-electronic, health, banking stock market and finance, defense industry, and optimization.

3.1 ANFIS

The main contribution of the fuzzy logic is that it provides calculation methodology which can deal with words, ambiguity, and level of detail. Human brain can interpret and process the incorrect and incomplete sensor information received from the organs. Similarly, fuzzy set theory can provide a systematic approach which deals with this information from a linguistic perspective. In addition, it can also make numerical calculations by using the membership function for the specified linguistic labels. The structure of fuzzy inference system (FIS) is based on fuzzy logic, fuzzy set theory, and fuzzy if-then rules. Framing of fuzzy if-then rules forms the key component of FIS. It is a very commonly used technique, and it is commonly applied in different fields such as system identification, automatic control, expert system, decision-making, model classification robotics, time series analysis, and forecasting. A fuzzy logic system consists of three basic components. These are a rule base, a data base, and a logic mechanism. Rule base consists of selected fuzzy rules, while a data base defines the membership functions of fuzzy rules. Fuzzy logic inference is modeled according to the rules to obtain a reasonable output or result. Jang introduced the adaptive network-based fuzzy inference system (ANFIS) by placing FIS within framework of customizable network. ANFIS is a neural fuzzy technique which is called adaptive network-based fuzzy inference system. ANFIS is also a neural fuzzy method in which the fusion process is carried out between the artificial neural network and fuzzy inference system. Its structure consists of a set of knots which are connected via adaptive network-oriented connections. The output of these adaptive knots depends on variable parameters related to these knots. Learning rule targets to minimize errors with update on these parameters. ANFIS methodology consists of a hybrid system which combines fuzzy logic and artificial neural network technique. While providing the artificial neural network with adaptation logic, fuzzy logic also considers the instability and ambiguity of the modeled system. By using this hybrid technique, together with the input variables, an initial fuzzy model structure is produced with the help of the rules gathered from the input and output data of the system. Afterward, neural network is used to produce the latest ANFIS model of the system and to regulate the rules of the first fuzzy model [45, 29].

3.1.1 ANFIS Structure

Basically, it is assumed that fuzzy inference system consists of two inputs and one output. The rule base includes the fuzzy if-then rules of Takagi and Sugeno. That is,

$$\text{If } x \text{ is } A,\ y \text{ is } B \text{ then } z \text{ is a funtion related to them } z = f\ (x, y)\,.$$

At this point x and y are the input variables, A and B are the fuzzy sets, and z is the output variable.

Generally, $f(x,y)$ is a polynomial for the x and y variables. But, there could also be a function serving any other purpose that is able to roughly describe the input of the system within this fuzzy section. If $f(x,y)$ has a fixed value, this forms the zero-degree Sugeno fuzzy model. This can be regarded as a special case of the Mamdani fuzzy inference system, in which each rule is indicated by a fuzzy uniqueness. If f(x,y) is a first-degree polynomial, this forms the first-degree Sugeno fuzzy model. Two rules for the first-degree Sugeno fuzzy inference system can be stated in this way:

$$\text{Rule 1 : If } x \text{ is A1 and } y \text{ is B1, then } f1 = \text{p1x} + \text{q1y} + \text{r1}$$

$$\text{Rule 2 : If } x \text{ is A2 and y is B2, then } f2 = \text{p2x} + \text{q2y} + \text{r2}$$

ANFIS structure is shown schematically in Fig. 7.

Figure 7 ANFIS structure in layered representation

Layers of ANFIS structure are described as follows:

Layer 1: Every knot in this layer is square and adaptive.

$$O_i^1 = \mu_{A_i}(x) \tag{4}$$

X is the input related to i knot. A_i is the linguistic variable related to knot function. $\mu_{Ai}(x)$ is the membership function of A_i. $\mu_{Ai}(x)$ can be chosen in two different ways. The first one is calculated in this way:

$$\mu_{A_i}(x) = \frac{1}{1 + \left[\left(\frac{x-c_i}{a_i}\right)^2\right]^{b_i}} \tag{5}$$

The second one is calculated in this way:

$$\mu_{A_i}(x) = \exp\left\{-\left(\frac{x - c_i}{a_i}\right)^2\right\} \tag{6}$$

At this point x is input. a_i, b_i, and c_i are precursor parameter sets.

Layer 2: Every knot in this layer is a fixed knot which calculates firing strength wi of a rule. The output of each knot is a product of all the signals it receives, and this is shown as follows:

$$O_i^2 = w_i = \mu_{A_i}(x) \times \mu_{B_i}(y),\ i = 1, 2 \tag{7}$$

Layer 3: Each knot in this layer has a fixed structure. Each i. knot calculates the ratio of the firing strength of i. rule to the combined firing powers of all the rules. i. output of i. knot shows the normalized firing level as shown below:

$$O_i^3 = \overline{w_i} = \frac{w_i}{w_1 + w_2},\ i = 1, 2 \tag{8}$$

Layer 4: Each knot in this layer is an adaptive knot which has a knot function as shown below:

$$O_i^4 = \overline{w_i} f_i = \overline{w_i}\left(p_i x + q_i y + r_i\right),\ i = 1, 2 \tag{9}$$

At this point $\overline{w}_i$ is the output of layer 3. p_i, q_i, and r_i are the outcome parameter sets.

Layer 5: This layer calculates the total output as a fixed knot which calculates it as the total of the signals coming from all the inputs.

$$O_i^5 = \sum_i \overline{\omega}_i f_i = \frac{\sum_i \omega_i f_i}{\sum_i \omega_i} \tag{10}$$

Considering the precursor parameter value, ANFIS structure can be expressed as linear combination of the resulting output parameters obtained as a result.

$$\begin{aligned} f &= \frac{w_1}{w_1+w_2} f_1 + \frac{w_2}{w_1+w_2} f_2 \\ &= \overline{w_1} f_1 + \overline{w_2} f_2 \\ &= (\overline{w_1} x)\, p_1 + (\overline{w_1} y)\, q_1 + (\overline{w_1})\, r_1 + (\overline{w_1} x)\, p_2 + (\overline{w_2} y)\, q_2 + (\overline{w_w})\, r_2 \end{aligned} \quad (11)$$

Here, f has a linear structure in p_i, q_i, and r_i parameters. This structure represents the learning algorithm of ANFIS. In the forward transition of the learning algorithm, the result parameters are defined by estimating the smallest squares, while in the backward transition, error signals spread out from the output layer to the input layer. In this backward transition, gradient descent algorithm is used to update the precursor parameters.

3.2 *OHS Risk Assessment in Underground Mining: A Case Study Using Multi-criteria Fuzzy Logic Approach*

The mine management tries to identify some of the most important hazards in the production process in order to take the necessary precautions and to prevent the incidence of occupational accidents. A team that consists of ten decision-makers (DM1–DM7) was set up to assess the types of hazards in five sections of the mine and the entire system. Decision-makers consist of seven mining engineers. Five types of major, totally 75 hazards are defined by the DM team as (H1–H75). Risk parameters are created as “possibility, severity, and sensitivity to nonuse of personal protective equipment and non-maintenance sensitivity and unpredictability” with the help of the relevant literature review and the views of the DM team (Chart 6).

Linguistic variables which are used by seven decision-makers are shown in Chart 7 in order to evaluate the subjective significance of the risk parameters. Linguistic rating of variables is shown in Chart 7 to assess the grading of hazard types according to each risk (Chart 8).

The Buckley FAHP method, expert’s assessment of linguistic variables, was used to calculate the subjective weights of the risk parameters through bilateral comparisons, and the results are given in Chart 9. The consistency rate calculated in this study was found to be less than 0.1 according to decision-makers’ calculations. Furthermore, the binary comparison matrix can be considered appropriate, and the questionnaire is valid in terms of FAHP (Fig. 5).

Chart 6 Underground mining risk groups

No	Risk groups
1	Dust
2	Unplanned power cut
3	Explosions
4	Chemical leakage
5	Ventilation system

Chart 7 Linguistic terms and related fuzzy values for risk parameters

AS	(2, 5/2, 3)	Absolutely strong
VS	(3/2,2, 5/2)	Very strong
FS	(1,3/2,2)	Fairly strong
SS	(1,1,3/2)	Slightly strong
E	(1, 1, 1)	Equal
SW	(2/3, 1, 1)	Slightly weak
FW	(1/2,2/3,1)	Fairly weak
VW	(2/5,1/2,2/3)	Very weak
AW	(1/3,2/5,1/2)	Absolutely weak

Chart 8 Linguistic terms and related fuzzy values for hazard assessment

VP	(0,0,1)	Very poor
P	(0,1,3)	Poor
MP	(1,3,5)	Medium poor
F	(3,5,7)	Fair
MG	(5,7,9)	Medium good
G	(7,9,10)	Good
VG	(9,10,10)	Very good

Chart 9 Buckley's FAHP risk parameters

Risk parameter	O			S			PPE			M			U		
O	1.0	1.0	1.0	0.5	0.7	0.9	0.5	0.7	0.9	0.7	1.0	1.0	0.7	0.9	1.2
S	1.2	1.4	1.9	1.0	1.0	1.0	0.8	1.0	1.3	0.8	1.3	1.5	0.7	1.1	1.5
PPE	1.1	1.5	2.0	0.8	1.0	1.2	1.0	1.0	1.0	1.0	1.0	1.5	1.0	1.0	1.5
M	1.0	1.0	1.5	0.7	0.8	1.2	0.7	1.0	1.0	1.0	1.0	1.0	0.8	1.0	1.2
U	0.8	1.1	1.4	0.7	0.9	1.3	0.7	1.0	1.0	0.8	1.0	1.3	1.0	1.0	1.0

O occurrence, *S* severity, *PPE* personal protective equipment, *M* sensitivity to maintenance, *U* undetectability

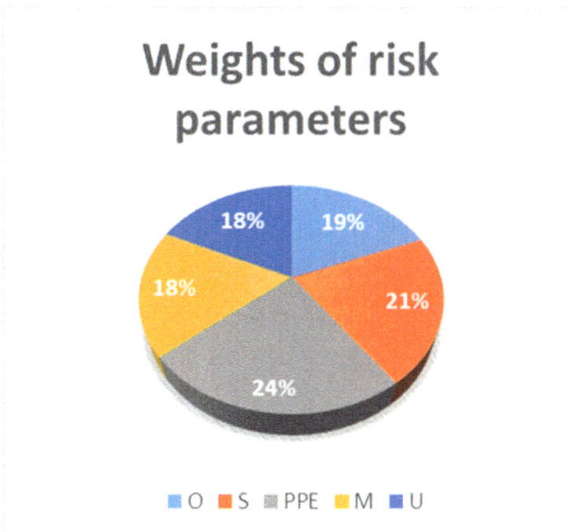

Fig. 5 Weights of risk parameters

Chart 10 S, R, and Q values for all hazard types of workplace

Hazards	H1	H2	H3	H4	H5	H6	H7	H8	H9	....	H75
Si	0.42	0.38	0.48	0.45	0.55	0.54	0.44	0.35	0.15	....	0.26
Ri	0.13	0.11	0.13	0.12	0.13	0.15	0.13	0.14	0.12	....	0.08
Qi	0.53	0.42	0.6	0.53	0.67	0.74	0.54	0.49	0.17	....	0.13

Chart 11 Risk groups in the workplace

Risk group
Dust
Unplanned power cut
Explosions
Chemical leakage
Ventilation system

After determination of the weight of five risk parameters, the fuzzy assessment of each risk parameter was applied with FVIKOR according to the hazard types in each of the six sections of the observed underground mine. In the study, the risk parameter evaluations of experts' linguistic variables were evaluated in total for all employees (i.e., "whole system" and "whole system evaluations of 75 different hazard types" have been carried out). Then, the resulting linguistic evaluations are transformed into triangular fuzzy numbers. Afterward, the cumulative fuzzy ratings of the hazard types are obtained to determine the fuzzy decision matrix. The fuzzy decision matrix for the mining workplace is given in Chart 10.

After all, the fuzzy best f_{j*} and fuzzy worst f_{j-} values of all risk parameter estimates are calculated using Eqs. (8) and (9). Normalized fuzzy distance is calculated for each risk parameter of the hazard types in the workplace. Afterward, the values of R, S, and Q have been calculated for all hazard types as in Chart 10. Finally, the risk precedence order of the danger types is determined according to S, R, and Q in decreasing order. While the minimum values have the highest risk rankings, the risks closest to 1 in S, R, and Q values represent the lowest risk.

From the obtained results, it can be understood that three most important hazard types in the observed mine are (H63), (H65), and (H13) orderly. The hazard types of (H57), (H58), (H63), and (H65) have been placed in the first rows in the observed mine on the ventilation system category. There is a total of five clusters as shown in Chart 11. Cluster 1 consists of five different hazard types; (H2) represents the hazard with the highest level of risk. Cluster 2 is composed of 7 different hazard types, Cluster 3 is composed 21 different hazard types, Cluster 4 is composed of 12 different hazard types, and finally Cluster 5 is composed of 31 different hazards.

4 Implementation

In this study, ANFIS-oriented model is implemented which will deal with underground mining risk analysis problem in three stages on the basis of neuro-fuzzy approach. Using the data gathered from a mine in Turkey which has over 1000

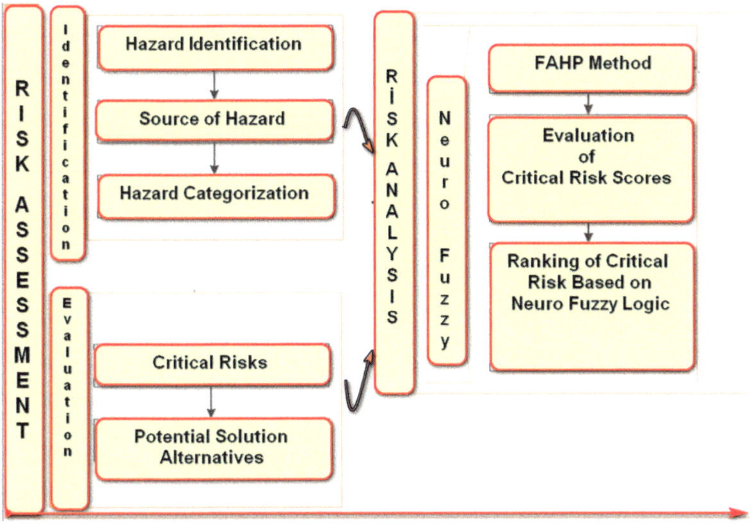

Fig. 6 Neuro-fuzzy logic and risk assessment model stages of OHS

workers with a capacity of over one million tones, risk assessment study was carried out based on neuro-fuzzy logic, and the comparison was made via classic risk assessment methods over error criterion.

In Fig. 6 the stages of the study carried out using neuro-fuzzy logic approach are given. In order to show the implemented method's productivity and flexibility, the obtained results will be compared via Buckley FAHP and FVIKOR two-stage risk assessment models.

Within Buckley FAHP method, there exist advantages, such as providing convenience in extension of fuzzy state and ensuring only one result. What is aimed within the implementation is to obtain a more efficient assessment with the total risk value as a measurable output by integrating neuro-fuzzy logic to risk analysis and obtain better results than classic risk assessment models.

Within the scope of methodology mentioned before, the inputs are specified as dust, unplanned power cut, explosions, chemical leakage, and ventilation system; the data set given in Appendix-A was obtained as a consequence of an elaborative cooperation by the specialists (mining engineer, occupational safety specialist, workplace doctor).

Within the frame of the first phase of the study, selection of the membership function is required to be processed with a convenient number and type for each criterion (input). Defining the number of the membership is one of the significant issues in the implementation of ANFIS model.

In case of selecting one membership function in relation of each input, the number of the parameter to be trained will be 84; for two membership functions, it will be 168; and for three membership functions, the number will be 252. Within the frame of OHS, in order to obtain a broader study, it was decided to select three

Table 1 Error values of membership functions

Membership function types	Fifty cycle error values
trimf	0.000013650
trapmf	0.000056420
gbellmf	0.000004395
gaussmf	0.000014573
gauss2mf	0.000004047
pimf	0.000004383
dsigmf	0.000008489
psigmf	0.000008653

membership functions. Followingly, the most suitable membership function for the model should be determined.

In the study's process for selecting the membership function types of the inputs, analyzing the result of error values which were received for 50 cycles, it was found that gauss2mf function type gave the least error value.

At this point, the data given in Appendix are named and uploaded as train data and test data.

Following the criteria selection, the problem changed to become neuro-fuzzy system with five inputs and one output. In order to define the number and membership types of the inputs, "Generate FIS" is used. Gauss2mf membership function is selected in the option of "MF Type." Output "MF Type" option should be selected as "linear." As mentioned previously, it represents the first-degree Sugeno neuro-fuzzy model (Table 1).

Once the data to be trained is uploaded to ANFIS editor and tested, information about the model can be found and the value which is "Epoch 50: error = 0,000004578." It means that for this selected membership function type and for each input, 3 triangle membership functions were trained for 50 cycles in accordance with hybrid learning algorithm, and the output value with an error value of 0,000004578 was forecasted accurately.

In ANFIS editor window structure, ANFIS model is shown (Fig. 7). On the very left side of the figure, the black nodes represent inputs; three separate second layer nodes which are connected to each node represent membership functions. There are five inputs, and each input has three membership functions that occur. The output received from the related rules is shown in the fifth layer; the risk value that the system forecasted in the fourth layer is shown in the fifth layer as the output value. In Fig. 8, the rules related to the established model are shown.

The rules descripted in the model can be edited by the specialist users within "Rule Editor." The first ten rules of risk assessment model can be observed in Fig. 8.

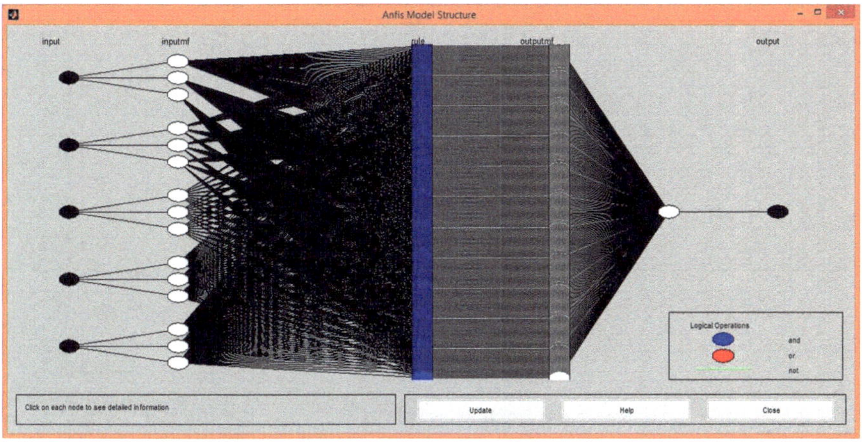

Fig. 7 Sample ANFIS model structure view

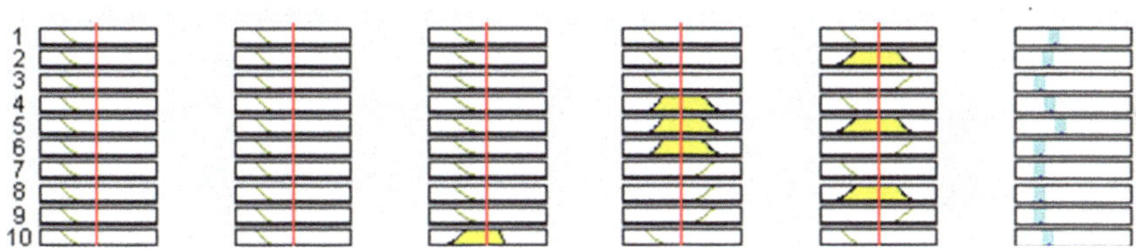

Fig. 8 The view of the first ten rules

It is possible to list 250 output values with the command of "evalfis" which is used in the command screen in Matlab. For this, the model should be defined and commanded as evalfis (input, model).

After building the data base which includes the risk groups for the underground mining risk analysis study, the effect of each input (risk group) over the total risk value used to be investigated. Since the risk value was expressed with a stochastic term, the process of normalizing in the obtained results was done via Excel. Due to the data used in the study and mining sector's general structure, error tolerance values are at very low levels. Hence, the datum may differ on percentile and permille basis. The first 180 is comprised of training data set, and the last 70 is comprised of test data set. First 18 training data and first 7 test data sample can be seen below (Table 2).

The way for calculating the risk groups' effect on the total risk values with ANFIS was investigated in OHS risk analysis study with neuro-fuzzy logic approach. The display of the inputs' fuzzy inference interface is obtained. The structure of the inputs can be edited via membership function editor. For each input three membership functions were assigned. Low, mid, and high verbal values are assigned for these three membership functions (Figs. 9 and 10).

Table 2 ANFIS model output values

Data type	Group	Risk value	ANFIS model
Training data	1	0.781	0.7029
	2	0.3105	0.27945
	3	0.81	0.729
	4	0.4689	0.42201
	5	0.8316	0.74844
	6	0.5626	0.50634
	7	0.67832	0.610488
	8	0.78208	0.703872
	9	0.45158	0.406422
	10	0.357	0.3213
	11	0.728832	0.655949
	12	0.724128	0.651715
	13	0.582072	0.523865
	14	0.571704	0.514534
	15	0.75378	0.678402
	16	0.621984	0.559786
	17	0.66683	0.600147
	18	0.690119	0.621107
Test data	19	0.713491	0.642142
	20	0.736946	0.663252
	21	0.760485	0.684437
	22	0.784108	0.705697
	23	0.807813	0.727032
	24	0.831602	0.748442
	25	0.855474	0.769927

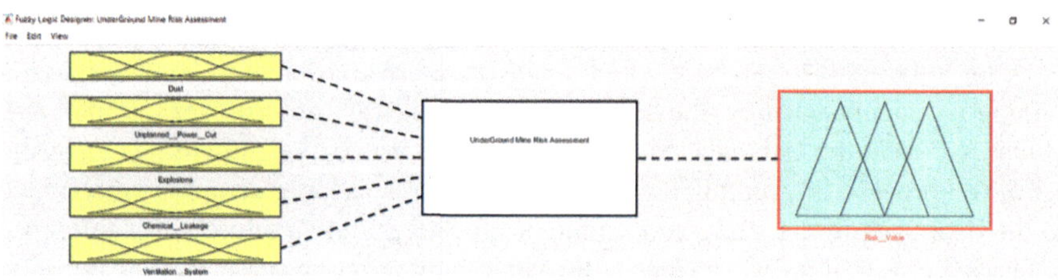

Fig. 9 The structure of model FIS

Risk assessment model includes 729 rules, and it comprises of 729 separate output types. For each output rule, six different verbal values are assigned as very low, medium low, low, medium high, high, and very high. The rule structure became meaningful after all the verbal value defining. The rules are displayed via "Rules" option located below "Edit" menu.

In this study, analysis of variables was carried out with the aim of observing the effect of the risk factors (dust, unplanned power cut, explosions, chemical

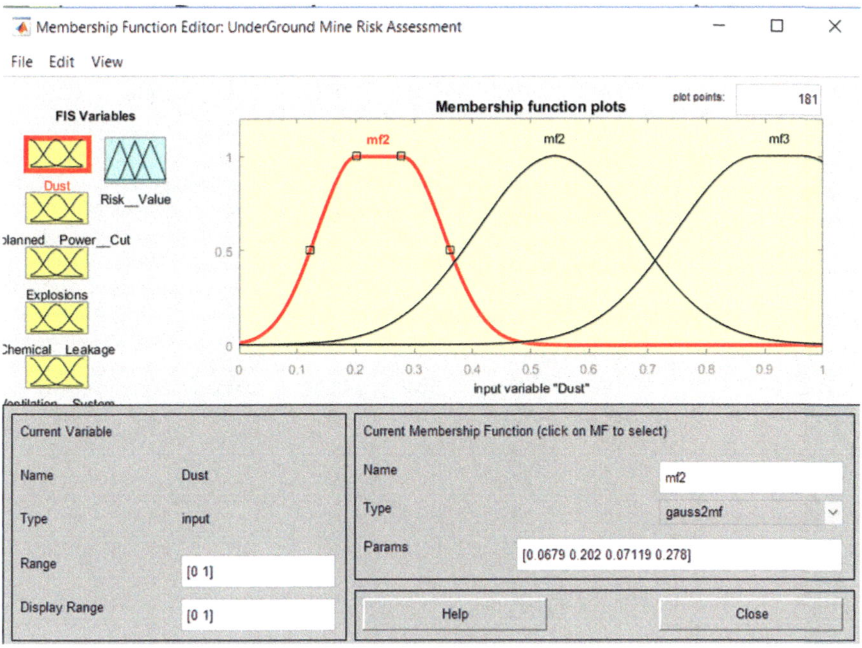

Fig. 10 The structures of input membership function

Table 3 Defining the criteria for dust risk

	Risk weight						
Risk group	DM1	DM2	DM3	DM4	DM5	DM6	DM7
Dust	MP	F	P	F	MP	G	P
Unplanned power cut	MG	G	F	MG	G	MG	F
Explosions	G	F	G	MG	F	MG	G
Chemical leakage	G	F	F	G	VG	G	VG
Ventilation system	MG	G	VG	G	VG	VG	G

leakage, and ventilation system) on general risk values. At this point, the effect of the inputs on the model result was observed by maintaining the model's rule number, membership function types, and membership function number stable. In Table 3, total risk value impact assessments are shown in terms of dust risk factor with decision-makers (DM), and the assessments were made previously.

In Table 4, impact assessments over total risk value are shown in terms of unplanned power cut risk factor.

In Table 5, impact assessments on total risk value are shown in terms of explosions risk factor.

In Table 6, impact assessments on total risk value are shown in terms of chemical leakage risk factor.

Table 4 The assessment of criteria for unplanned power cut risk

	Risk weight						
Risk group	DM1	DM2	DM3	DM4	DM5	DM6	DM7
Dust	MP	MG	MP	G	G	G	MP
Unplanned power cut	MG	G	F	MG	G	F	G
Explosions	G	F	G	MP	G	MG	G
Chemical leakage	G	MG	VG	VG	MG	F	G
Ventilation system	G	G	MG	G	F	VG	MG

Table 5 The assessment of criteria for explosions risk

	Risk weight						
Risk group	DM1	DM2	DM3	DM4	DM5	DM6	DM7
Dust	MG	G	F	MG	MG	MG	MP
Unplanned power cut	F	MG	MG	MP	VG	G	MG
Explosions	MG	MG	G	MG	F	G	VG
Chemical leakage	G	VG	F	G	F	MG	G
Ventilation system	F	VG	F	MG	MP	F	VG

Table 6 The assessment of criteria for chemical leakage risk

	Risk weight						
Risk group	DM1	DM2	DM3	DM4	DM5	DM6	DM7
Dust	P	G	F	F	G	P	F
Unplanned power cut	MP	P	G	G	G	MG	G
Explosions	F	F	G	MP	VG	G	MP
Chemical leakage	G	G	F	MP	F	VG	G
Ventilation system	F	VG	G	MG	G	G	VG

Table 7 The assessment of criteria for ventilation system risk

	Risk weight						
Risk group	DM1	DM2	DM3	DM4	DM5	DM6	DM7
Dust	MG	G	MG	F	G	G	G
Unplanned power cut	VG	MG	G	G	MG	F	G
Explosions	G	VG	MG	MG	F	MG	G
Chemical leakage	MG	G	F	G	VG	G	VG
Ventilation system	VG	F	VG	G	MG	G	VG

In Table 7, impact assessments on total risk value are shown in terms of ventilation system risk factor.

In accordance with these assessments, the sensitivity of ANFIS on risk factors was tested, and the total risk value range results based on the factors are shown in Table 8. Accordingly, the following result was found: any change in ventilation system risk group as the highest active change can show an impact on total risk value

Table 8 Total risk value ranges by factors

Risk group	Risk value range
Ventilation system	0–0,124
Explosions	0–0,082
Chemical leakage	0–0,077
Unplanned power cut	0–0,049
Dust	0–0,034

by 0,124; similarly, any change in explosions risk group can show an impact on total risk value by 0,082; in chemical leakage risk group, it can show an impact on total risk value by 0,077; in unplanned power cut risk group, it can show an impact on total risk value by 0,049; and in dust risk group, it can show an impact on total risk value by 0,034.

5 Results and Recommendations

In underground mining risk assessment, after defining the hazard groups, since gauss2mf membership function gave the lowest RMSE level results, this function was selected and commenced. The model which was assigned and developed in the beginning was consisted of 5 inputs, 1 output, and 729 rules. The results were obtained and normalized. In accordance with the results, the precautions to be taken were formed by the specialists. Training data set is consisted of 18 data, and the test data set is consisted of 7 data.

One of the most significant elements in OHS is risk assessment. Taking the study about the risk assessment in literature in account and risk assessment is a multi-criteria decision-making problem which tackles risk value optimization and includes many criteria. In this study, risk assessment model was obtained by the studies and gap on them in literature, underground mine structure, and also specialist views. In this underground mining-based risk assessment study, after defining the mining and department-based risk groups and hazard groups, the significant order of each risk group was defined, and the data set used in risk assessment with the learning feature of ANFIS and its adaptive network-based structure resulted with less error compared to classic methods. The suggested risk assessment model can be used in district-, province-, region-, and country-wide risk assessment stages. The established model with the need of degrading the coefficient of risk of the mines which is among the most significant units of underground mining sector. It is consisted of five total inputs which have three membership functions and seven outputs. In when it will be used bigger data set, the process could take longer. In order to compare the superiority of the fuzzy logic approach on classic and fuzzy risk assessments in the presented OHS, the error values of FAHP and L-matrix output results were compared over the same data set (Table 9).

Table 9 Neuro-fuzzy logic-based approach result comparison

	Training data set		Test data set	
Model	MSE	RMSE	MSE	RMSE
ANFIS	0,057	0,239	0,091	0,302
FAHP	0,076	0,276	0,124	0,352
L-matrix	0,107	0,327	0,165	0,406

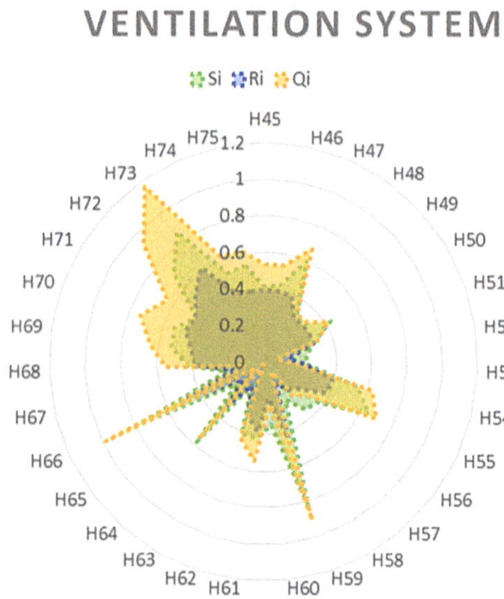

Fig. 11 Risk priorities of hazard types on ventilation system

As a result, it was found that the most significant risk group of the observed mine's whole system is ventilation system risk group and it stems from the assembly and disassembly of fans, working at height; assembly and disassembly of fans, suspension; and assembly and disassembly of fans, load lifting in itself. Explosions is the second most significant risk group, and it consists of hazards such as spreading of acidic water around, high slope, contact with acidic water in or around the sulfurous tallow, vehicle crash, damper tipping, and spilling of sulfur material on the way of transport. Chemical leakage is the third highest risk group which consists of hazards such as acidic environment, electric shock, temperature, quality of visibility, and SO2 formation. Unplanned power cut risk group is following chemical leakage with hazards such as paste fill pipeline clogging, being stuck in an elevator, stopping of the pumps, stopping of compressors, rescue chamber energy and air cutoff, cement working equipment, and stopping of the fans. Risk group which has the lowest risk score is dust with hazards of occupational disease, quality of visibility, breakdown of vehicles, and dust explosion. Figures 11, 12, 13, 14, and 15 show the priorities of each risk group.

The risk priority orders of the hazard types are obtained with parameters R,S, and Q. There is a negative correlation between parameters and risk significance. While the total value of parameters approaches to 0, corresponding risk ranked as highest risk. After obtaining risk priorities for each group, compromised risk rankings are

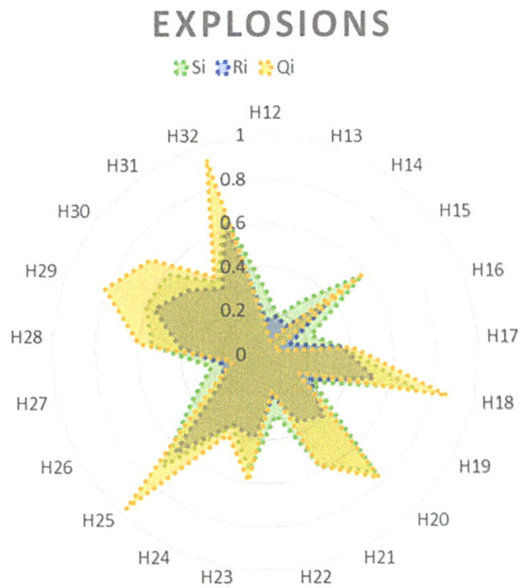

Fig. 12 Risk priorities of hazard types on explosions

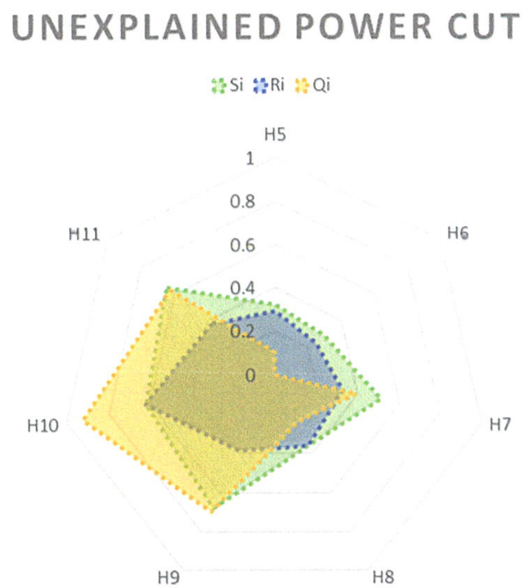

Fig. 13 Risk priorities of hazard types on unexplained power cut

found. Figures 16, 17, 18, 19, and 20 show the compromised rankings of hazards with respect to their groups.

After defining the risk groups' significant order, the precautions to be taken should be defined accordingly. For this, neuro-fuzzy logic results were assessed by the specialists and the precautions to be taken. The reason behind this idea is elimination of hazards, checking the hazards in their sources, minimization of hazards, and providing the most convenient PPE.

Department-based precautions to be taken are defined. For the ventilation system, giving an advanced mine rescue training, determination of appropriate location,

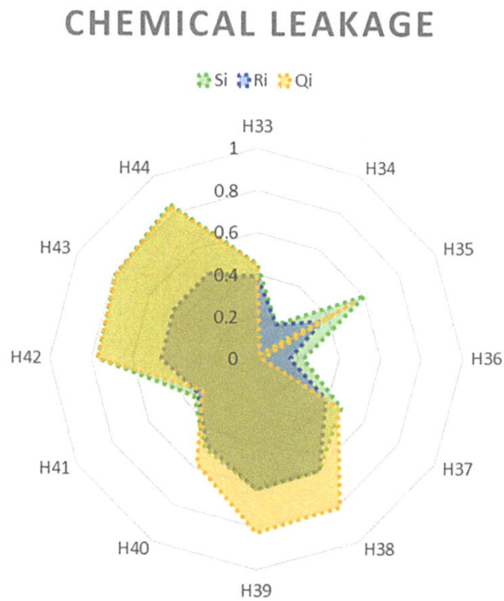

Fig. 14 Risk priorities of hazard types on chemical leakage

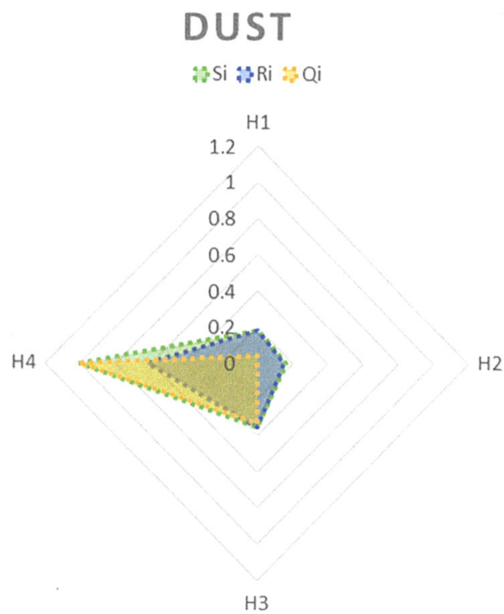

Fig. 15 Risk priorities of hazard types on dust

providing natural ventilation gas conversion structure, periodically and regular examination and improvement on transportation vehicles, procedure for working in high up, following predefined ventilation standards, availability of opened roof of the elevator from inside, providing clean air outlets, temporary lane closures with signs for workplace, providing personal escape mask, pressure tubes, giving an advanced education and authorization before works, carrying out hot work permit

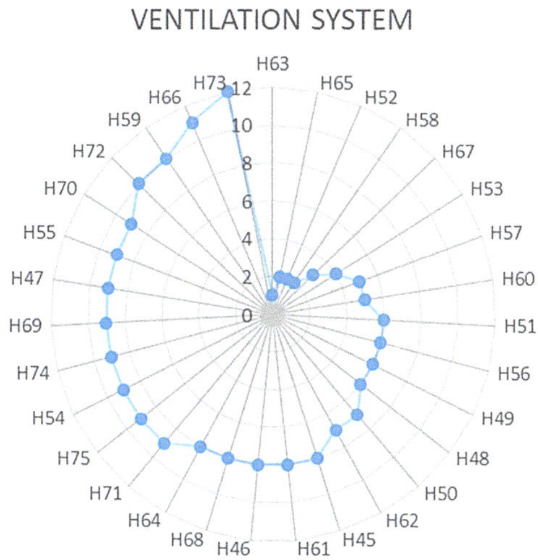

Fig. 16 Compromised rankings for the hazard types on ventilation system

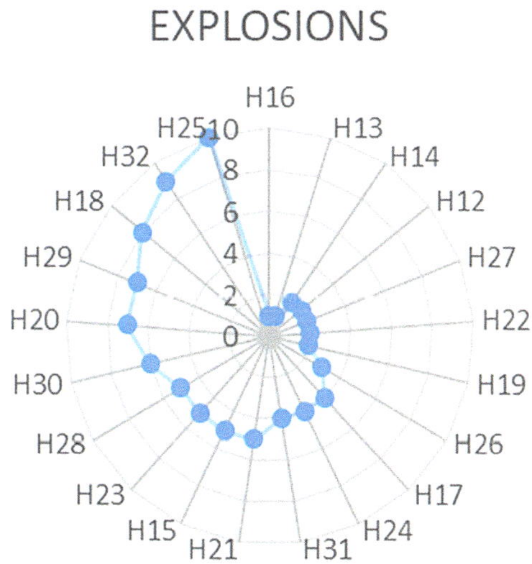

Fig. 17 Compromised rankings for the hazard types on explosions

form, regular checks of air quality with appropriate sensors, preferring experienced mine operators and providing appropriate PPE, immobilization of equipment during transportation, regular controls, providing proper fortification standards, periodic maintenance and propitiously fixing of vehicles, providing safety lamps, filling individual identification number, and preparing general workplace inspection checklist are the major control measures with respect to occupational safety and health. For the explosions, appropriate controls and cleaning of miss-fires, providing proper

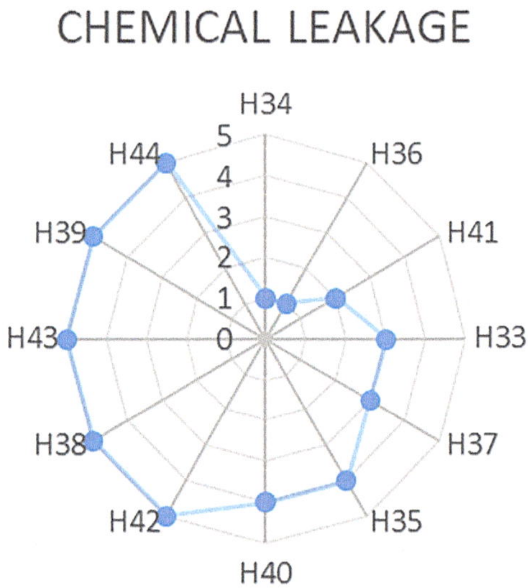

Fig. 18 Compromised rankings for the hazard types on chemical leakage

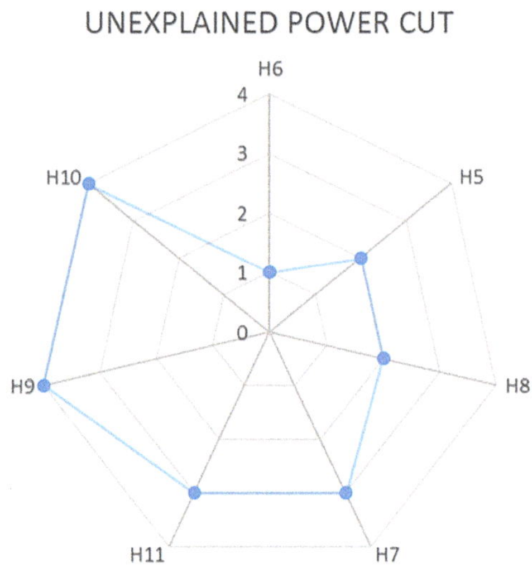

Fig. 19 Compromised rankings for the hazard types on unexplained power cut

fortification standards, blasting checks, training of driving to avoid crashes, giving an advanced training before works, eliminating the more dangerous explosives with less dangerous explosives to decrease risk level, preferring well-educated and certificated workers, and regular maintenance of equipment, separate transport of explosives and capsules, international lightning protection codes and standards, investigation of the geological structure of the area before explosions, providing

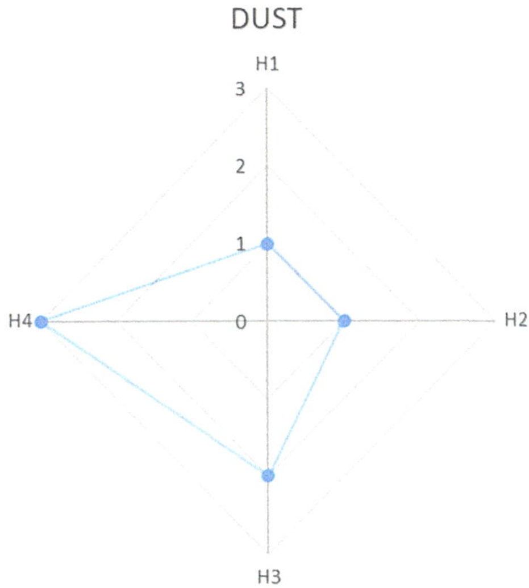

Fig. 20 Compromised rankings for the hazard types on dust

of proper PPE, design of technical surveillance, filling individual identification number, providing safety barricading procedure, notification of all workers before explosion by mine control, building barricade start-up checklist, signal system for key points building hot work permit form, providing an explosive management plan and microseismic monitoring system, pre-work controls, obligation for vocational education, providing rock mechanics testing equipment, providing proper fortification standards that are control measures for continuous paste fill activity, remote distance control system regulations for ventilation and air-conditioned equipment, control of hazardous energy with procedure, application of job safety analysis, and preparation for compressed air and pressurized waters are necessary control measures. Chemical leakage requires following control measures, providing proper PPE procedure for transporting explosives, set up radio and camera communication system, avoiding flammable materials in transportation, installation requirements for gas and electric structure, chemical spill procedure filling individual identification number, preparation procedure of chemicals, controlling all system before starting of activity, authorized and experienced personal, appropriate vehicles, regular and systematic controls of work area, providing automatic fire suppression system, follow-up legislation, providing methodical mine rescue fire training, periodic health checks, selection and use of gloves procedure, occupational hazard analysis periodically, periodic pull test, advanced education before works, preference of cut-resistant gloves, material handling training, providing automatic fire suppression system, using specific checklist (oxygen set), periodic pull test, building hot work permit, checklist for battery voltage control panel, providing backup power lines, set up proper and effective air distribution, giving an advanced education and authorization before works, natural ventilation and radio communication system, availability of opened roof of the elevator from inside, periodic controls and maintenance of cables and panels (weekly/monthly/yearly), adequate number of safety switches, usage of the isolated cable and earth leakage circuit, grounding system, working under low voltage (24 volts), grounding of electrical panels properly, usage of safety hazard warning signs, gate transition procedure, energy insulation and locking procedure, the use of a double door system and one of them should permanently close mine rescue chamber, and previous common suggestions are major control measures for unplanned power cut risk. Speed limit enforcement (3.8 m/s), authorization and advanced education before works, follow-up legislation, surface topography measurements, providing sui and ergonomic PPE, regular cleaning of dusty areas, dust suppression system, periodic supervisory checks, usage of powder vacuum tool and main fan filter system, ambient and particular dust measurements, proper vehicle lightings, preference of reflective dress, improvement of vehicle visibility, periodic supervisory checks, washing equipment and vehicles periodically, periodic and regular cleaning on truck roads with broom and bucket, changing and improving vehicle exhaust system periodically, and cleaning mirrors and stopes are major control measures for dust.

Appendix

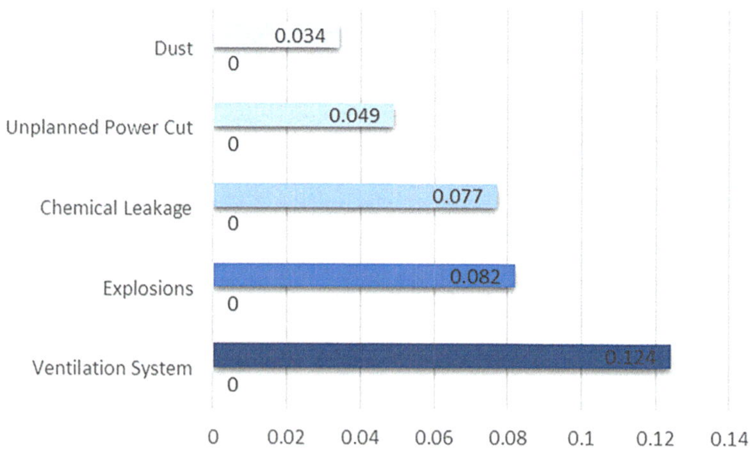

Fig. 21 Total risk value ranking by order

Chart 12 Risk groups and their hazard list

Risk group	Hazard ID	Hazard
Dust	H1	Occupational disease
	H2	Quality of visibility
	H3	Breakdown of vehicles
	H4	Dust explosion
Unplanned power cut	H5	Paste fill pipeline clogging
	H6	Being stuck in an elevator
	H7	Stopping of the pumps
	H8	Stopping of compressors
	H9	Rescue chamber energy and air cutoff
	H10	Cement working equipment
	H11	Stopping of the fans
Explosions	H12	Fire
	H13	Pressurized gas
	H14	Poisonous gas
	H15	Spark formation
	H16	Eye deterioration
	H17	Explosion
	H18	Vehicle accident
	H19	Fire
	H20	Dropping of the explosives from the vehicle
	H21	Electrical short circuit in the equipment
	H22	Sabotage
	H23	Stolen explosive material
	H24	Static electricity
	H25	Stroke of lightning
	H26	Spreading of acidic water around
	H27	High slope
	H28	Contact with acidic water in or around the sulfurous tallow
	H29	Vehicle crash
	H30	Damper tipping
	H31	Spilling of sulfur material on the way of transport
	H32	Uncovering working environment
Chemical leakage	H33	SO2 gas
	H34	Temperature
	H35	Quality of visibility
	H36	Acidic environment
	H37	SO2 formation
	H38	Unexploded explosives
	H39	Fall from height
	H40	Hand jamming
	H41	Electric shock
	H42	Hose burst
	H43	Penetration of energy line

(continued)

Chart 12 (continued)

Risk group	Hazard ID	Hazard
	H44	Damages of rot
Ventilation system	H45	Hitting of ventilation doors to people
	H46	Compression of pistons
	H47	Blowing of employees working through the doorway by ventilation air
	H48	Hitting of vehicles to ventilation doors
	H49	Electric shock
	H50	Determination of fan location: Being not appropriate in terms of ground support
	H51	Determination of fan location: Unsuitability of drift section
	H52	Determination of fan location: Water tunnel
	H53	Nailing bolts to install fan: Do not nail a suitable bolt
	H54	Nailing bolts to install fan: Do not nail bolts in appropriate pattern
	H55	Loading and unloading of fans for transport: Suspended fan
	H56	Loading and unloading of fans for transport: Falling of fan from height
	H57	Loading and unloading of fans for transport: Wrong bearing element selection
	H58	Loading and unloading of fans for transport: Lifting equipment
	H59	Loading and unloading of fans for transport: Authorization
	H60	Transport of fans: Transporter
	H61	Transport of fans: Fixing the fan
	H62	Transport of fans: Inappropriate loading of the fan
	H63	Assembly and disassembly of fans: Working at height
	H64	Assembly and disassembly of fans: Load lifting
	H65	Assembly and disassembly of fans: Suspension
	H66	Assembly and disassembly of fans: Working at narrow area
	H67	Assembly and disassembly of fans: Assembly elements
	H68	Assembly and disassembly of fans: Ventilation
	H69	Assembly and disassembly of fans: Uncontrolled movement of fan
	H70	Assembly and disassembly of fans: Hot works
	H71	Engaging the fan: Diffuser and adapter selection
	H72	Engaging the fan: Electricity
	H73	Engaging the fan: Working at height
	H74	Engaging the fan: Working at narrow area
	H75	Periodic maintenance and control of fans: Corrosion

References

1. Al-Turjman, F. (2017). Price-based data delivery framework for dynamic and pervasive IoT. *Elsevier Pervasive and Mobile Computing Journal, 42*, 299–316.
2. Al-Turjman, F. (2017). Energy–aware data delivery framework for safety-oriented mobile IoT. *IEEE Sensors Journal, 18*(1), 470–478.
3. Singh, A., Kumar, D., & Hötzel, J. (2018). IoT Based information and communication system for enhancing underground mines safety and productivity: Genesis, taxonomy and open issues. *Ad Hoc Networks, 78*, 115–129.
4. Al-Turjman, F., Hassanein, H., & Ibnkahla, M. (2013). Efficient deployment of wireless sensor networks targeting environment monitoring applications. *Elsevier: Computer Communications Journal, 36*(2), 135–148.
5. Al-Turjman, F., Hassanein, H., & Ibnkahla, M. (2015). Towards prolonged lifetime for deployed WSNs in outdoor environment monitoring. *Elsevier Ad Hoc Networks Journal, 24*(A), 172–185.
6. Singh, G., & Al-Turjman, F. (2016). A data delivery framework for cognitive information-centric sensor networks in smart outdoor monitoring. *Elsevier Computer Communications Journal, 74*(1), 38–51.
7. Chan, H. K., & Wang, X. (2013). Fuzzy extent analysis for food risk assessment. In *Fuzzy Hierarchical Model for Risk Assessment* (pp. 89–114). London: Springer.
8. Ottawa Charter for Health Promotion. (1986). WHO/HPR/HEP/95.1. WHO, Geneva.
9. Chang, D. Y. (1996). Applications of the extent analysis method on fuzzy AHP. *European Journal of Operational Research, 95*(3), 649–655.
10. Djapan, M. J., Tadic, D. P., Macuzic, I. D., & Dragojovic, P. D. (2015). A new fuzzy model for determining risk level on the workplaces in manufacturing small and medium enterprises. *Proceedings of the Institution of Mechanical Engineers, Part O: Journal of Risk and Reliability, 229*(5), 456–468.
11. Buckley, J. J. (1985). Fuzzy hierarchical analysis. *Fuzzy Sets and Systems, 17*(3), 233–247.
12. Ebrahimnejad, S., Mousavi, S. M., & Seyrafianpour, H. (2010). Risk identification and assessment for build–operate–transfer projects: A fuzzy multi attribute decision making model. *Expert Systems with Applications, 37*(1), 575–586.
13. Aragonés-Beltrán, P., Mendoza-Roca, J. A., Bes-Piá, A., García-Melón, M., & Parra-Ruiz, E. (2009). Application of multi criteria decision analysis to jar-test results for chemicals selection in the physical–chemical treatment of textile wastewater. *Journal of Hazardous Materials, 164*(1), 288–295.
14. Cooper, M. D. (2000). Towards a model of safety culture. *Safety Science, 36*, s.113.
15. Gorman, T., Dropkin, J., Kamen, J., Nimbalkar, S., Zuckerman, N., Lowe, T., Szeinuk, J., Milek, D., Piligian, G., & Freund, A. (2013). Controlling health hazards to hospital workers. *New Solutions, 23*, 1–67.
16. Grassi, A., Gamberini, R., Mora, C., & Rimini, B. (2009). A fuzzy multi-attribute model for risk evaluation in workplaces. *Safety Science, 47*(5), 707–716.
17. Gul, M., & Ak, M. F. (2018). A comparative outline for quantifying risk ratings in occupational health and safety risk assessment. *Journal of Cleaner Production, 196*, 653–664.
18. Gul, M., & Guneri, A. F. (2016). A fuzzy multi criteria risk assessment based on decision matrix technique: A case study for aluminum industry. *Journal of Loss Prevention in the Process Industries, 40*, 89–100.
19. Gul, M., Celik, E., Aydin, N., Gumus, A. T., & Guneri, A. F. (2016). A state of the art literature review of VIKOR and its fuzzy extensions on applications. *Applied Soft Computing, 46*, 60–89.
20. Guneri, A. F., Gul, M., & Ozgurler, S. (2015). A fuzzy AHP methodology for selection of risk assessment methods in occupational safety. *International Journal of Risk Assessment and Management, 18*(3–4), 319–335.
21. Hu, A. H., Hsu, C. W., Kuo, T. C., & Wu, W. C. (2009). Risk evaluation of green components to hazardous substance using FMEA and FAHP. *Expert Systems with Applications, 36*(3), 7142–7147.

22. Jamshidi, A., Rahimi, S. A., Ait-Kadi, D., & Ruiz, A. (2015). A comprehensive fuzzy risk-based maintenance framework for prioritization of medical devices. *Applied Soft Computing, 32*, 322–334.
23. John, A., Paraskevadakis, D., Bury, A., Yang, Z., Riahi, R., & Wang, J. (2014). An integrated fuzzy risk assessment for seaport operations. *Safety Science, 68*, 180–194.
24. Kenya Ministries of Health and IntraHealth International. (2013). Report of the occupational safety and health risk assessment. Nairobi, Kenya: MsOH.
25. Liu, H. C., Fan, X. J., Li, P., & Chen, Y. Z. (2014). Evaluating the risk of failure modes with extended MULTIMOORA method under fuzzy environment. *Engineering Applications of Artificial Intelligence, 34*, 168–177.
26. Liu, H. C., Liu, L., Liu, N., & Mao, L. X. (2012). Risk evaluation in failure mode and effects analysis with extended VIKOR method under fuzzy environment. *Expert Systems with Applications, 39*(17), 12926–12934.
27. Liu, H. C., You, J. X., You, X. Y., & Shan, M. M. (2015). A novel approach for failure mode and effects analysis using combination weighting and fuzzy VIKOR method. *Applied Soft Computing, 28*, 579–588.
28. Liu, H. T., & Tsai, Y. L. (2012). A fuzzy risk assessment approach for occupational hazards in the construction industry. *Safety Science, 50*(4), 1067–1078.
29. Mahdevari, S., Shahriar, K., & Esfahanipour, A. (2014). Human health and safety risks management in underground coal mines using fuzzy TOPSIS. *Science of the Total Environment, 488*, 85–99.
30. Opricovic, S. (1998). Multicriteria optimization of civil engineering systems. *Faculty of Civil Engineering, Belgrade, 2*(1), 5–21.
31. Saaty, T. L. (1990). How to make a decision: The analytic hierarchy process. *European Journal of Operational Research, 48*(1), 9–26.
32. Tzeng, G., & Huang, J. (2011). *Multiple attribute decision making: Methods and applications.* Boca Raton, FL: Chapman & Hall/CRC.
33. Vahdani, B., Salimi, M., & Charkhchian, M. (2015). A new FMEA method by integrating fuzzy belief structure and TOPSIS to improve risk evaluation process. *The International Journal of Advanced Manufacturing Technology, 77*(1–4), 357–368.
34. Amalberti, R. (2001). The paradoxes of almost totally safe transportation systems. *Safety Science., 37*, 109–126.
35. Lin, P. H. (2001). *Safety management and risk modelling in aviation – The challenge of quantifying management influences*, PhD thesis, Delft University of Technology.
36. Xianfeng, L., & Shengguo, H. (2012). Airport safety risk evaluation based on modification of quantitative safety management model. *Procedia Engineering., 43*, 238–244.
37. Elgstrand, K., & Vingärd, E. (2013). *Occupational safety and health in mining: Anthology on the situation in 16 mining countries.* Gotenburg: University of Gothenburg.
38. Verma, S., & Chaudhri, S. (2014). Integration of fuzzy reasoning approach (FRA) and fuzzy analytic hierarchy process (FAHP) for risk assessment in mining industry. *Journal of Industrial Engineering and Management, 7*(5), 1347.
39. Aven, T. (2009). *Risk analysis: Assessing uncertainties beyond expected values and probabilities.* Chichester: Wiley.
40. Awasthi, A., & Kannan, G. (2016). Green supplier development program selection using NGT and VIKOR under fuzzy environment. *Computers & Industrial Engineering, 91*, 100–108.
41. Badri, A. (2015). The challenge of integrating OHS into industrial project risk management: Proposal of a methodological approach to guide future research (case of mining projects in Quebec, Canada). *Minerals, 5*(2), 314–334.
42. Badri, A., Nadeau, S., & Gbodossou, A. (2013). A new practical approach to risk management for underground mining project in Quebec. *Journal of Loss Prevention in the Process Industries, 26*(6), 1145–1158.
43. Chen, C. T. (2000). Extensions of the TOPSIS for group decision-making under fuzzy environment. *Fuzzy Sets and Systems, 114*(1), 1–9.

44. Donoghue, A. M. (2004). Occupational health hazards in mining: An overview. *Occupational Medicine, 54*(5), 283–289.
45. Lang, L., & Fu-Bao, Z. (2010). A comprehensive hazard evaluation system for spontaneous combustion of coal in underground mining. *International Journal of Coal Geology, 82*(1), 27–36.

Intelligent IoT Communication in Smart Environments: An Overview

Joel Poncha Lemayian and Fadi Al-Turjman

1 Introduction

Presently, it is estimated that more than half of the world's population live in the urban areas. The massive shift from rural to urban areas is projected to continue in the next few decades. Figure 1 shows the comparison of the population living in the rural versus urban places from the year 1950 to 2050 [2]. The author in [5] states that it is projected that 70% of the world's population will be living in urban areas by the year 2050. This rapid migration is guaranteed to bring about chaos and disorder in the existing destination cities; therefore, we must adapt to the change and create better settlement areas to accommodate more people. Some of the challenges brought about by the rapid population increase in the urban area include waste management problems, scarcity of resources, air pollution, human health concerns, traffic congestions, and scarce, failing, and aging infrastructures, among others. "Smart" cities have emerged as one promising way of solving and controlling the problems faced due to increased population in cities. The author in [5] says that ICT (information and communication technologies) provides the cities with the ability to become "smart," whereas, it can efficiently manage city services. Nevertheless, researchers today have not given this topic much thought, as it supposed to be. Depending on the capital invested, the available land for construction and the technology present and the size and layout of a smart city can vary from small shopping malls to massive multimillion cities. For example, the government of Singapore has agreed to construct a smart city in India, costing about £10.7bn ($16.5bn) [2]. The city has an area of 7235 square kilometers, which is about ten times the size of their own 726-square-kilometer city. Some of the

J. P. Lemayian · F. Al-Turjman (✉)
Department of Computer Engineering, Antalya Bilim University, Antalya, Turkey
e-mail: lemayian.joel@std.antalya.edu.tr; fadi.alturjman@antalya.edu.tr

F. Al-Turjman (ed.), *Artificial Intelligence in IoT*,
Transactions on Computational Science and Computational Intelligence,
https://doi.org/10.1007/978-3-030-04110-6_10

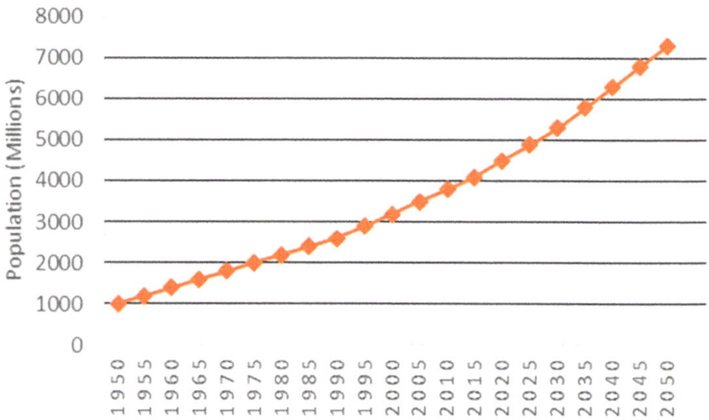

Fig. 1 Urban and rural population of the world, 1950–2050 [2]

countries that have truly embraced the development of smart cities include the USA, which has the highest number of smart cities according to IoT analytics, South Korea, and Germany.

The "smart" paradigm in smart cities is highly motivated by the ability of a city to be environmentally friendly; cities are edging toward conserving the environment by producing less waste that can pollute the environment and, hence, becoming green and environmentally sustainable [40]. Consequently, these cities are mostly put up in a fairly flat, clear areas, where there are few trees to be cleared to make way for buildings and where the area is easily accessible so that construction materials can easily be transported. Majority of smart cities are made up of high-rise buildings, because these buildings are able to accommodate many people in a fairly small land area. These buildings can either be residential, commercial, or both. There is a massive amount of data transmitted back and forth to a base station, from sensors located all over a smart city (smart parking system, smart traffic light systems, etc.) [41, 43]. The number of base stations and the distance between them is proportional to the size of the smart city; if a city is very big, various substations might need to be set up in order to ensure a safe and fast data transmission system. Moreover, the amount of data transmitted and the distance between transmitters and receivers determine the strength of the signals. Cybersecurity is one of the major challenges faced by smart systems at the moment. As a smart system, smart cities incorporate things like vehicles, homes, public venues, and other social systems which are on their way to gaining full connectivity through the IoT (Internet of Things) paradigm [38, 48]. Therefore, it is critical that both private and public information remain as secure as possible. Another major obstacle that is challenging the development of smart cities is power. Smart cities mostly use solar power for electricity production [33]; however, power production using solar panels is not always reliable, as it is affected by weather conditions such as changing seasons. It is therefore imperative that other forms of green power production are explored so as to power smart cities, for instance, the use of smart solar microgrids using Zigbee suggested by authors in [45].

However, smart cities are more than fast Internet connection, big data transmission, and interlinked applications; rather, the key is to set humans at the core of the smart solution, so as to make the most of the smart technology. As previously stated, several countries around the world have invested in smart cities, and in some places, funding has been done by private organizations. In Africa as well, there are several smart cities springing up in different countries; in particular, South Africa, Egypt, and Kenya are among some of the promising grounds where private organizations have come in to invest in the construction of smart cities.

The rest of this work is organized as follows. Section 2 gives an overview of the smart cities around the world. Section 3 gives the structure of a typical smart city, Sect. 4 looks at the flow of information in a smart city, while Sect. 5 talks about the most preferred form of communication system in a smart city. Finally, Sect. 6 concludes the study.

2 Smart Cities Around the World

In this section, we briefly look at some of the top smart cities in different parts of the world. The authors in [1] talk about smart cities around the world, focusing on smart cities found in America, South Korea, and Germany. According to IoT analytics, seven of the top ten IoT smart cities are found in America, with San Francisco taking the first position [1]. According to the authors, San Francisco is a host to several small-scale and large-scale IoT companies, including Cisco, Google, Apple, and Intel. However, most of the other cities in the USA lack smart city infrastructures such as ICT infrastructure, for instance, only 7.7% of Americans have access to optical fiber Internet, the fastest and the highest quality of Internet connectivity, showing that there is still room for development, even in the most developed countries. However, in a Smart Cities Initiative released in September 2015, President Obama administration recognized the great potential of smart cities in America. This led to the allocation of $ 45 million by the National Science Foundation and National Institute of Standards and Technology to build a research center for smart cities. Additionally, an extra $ 115 million was invested to finding solutions to public policy challenges. Moreover, the current government has made investment grands totaling about $ 3.4 billion toward smart grid development; this proves that the current government is determined to maintain the current position of the USA in terms of smart city development. Given that US smart city model represents the most market-driven approach of smart cities, it will be very interesting to see the kind of solutions and business models that will be present in a few years. This will certainly have a profound impact on the perception and development of smart cities around the world.

South Korea is one other country with massive smart city development projects in the world; the country has maintained the top position at the United Nations E-Government Development Index since 2003, and this is due to an impeccable ICT infrastructure among other things. In South Korea, there is an escalated use of ultrafast LTE network (4G), making it the most connected city in the world. For instance, in Seoul alone, the local government has sponsored over 831 free Wi-Fi zones; moreover, a major bank in South Korea has funded the installation of chargers at the free Wi-Fi zones, so that everyone can charge their phones while using the Internet. The government of South Korea has recently agreed to upgrade the fourth generation network used in the cities to fifth generation network by the year 2020; this will make the Internet connectivity 1000 times faster. Moreover, in 2014, the European Union backed up the government of South Korea and agreed to be part of developing the ultrafast fifth-generation wireless communications networks (5G) in.

The backbone of smart city development in Germany is sustainable growth, transportation, security, data protection, and how smart solutions help improve energy management to bring about energy efficiency. According to [1], in 2010, the German government has started a project to phase out the use of nuclear power; this creates room for the invention of better, greener, and more efficient smart energy sources. Energy efficiency is one of the driving factors that is causing Germany to develop smarter ways of power production. For instance, over 140 counties, municipalities, regional associations, and cities in Germany are part of a project called "100 % Erneuerbare-Energie-Regionen." This project identifies regions in the country that want to convert all their future energy sources to renewable energy. The project supports dedicated regions through communication, transfer, and networking services. Figure 2 shows some of the smartest cities in Europe.

Fig. 2 Smart cities in Europe

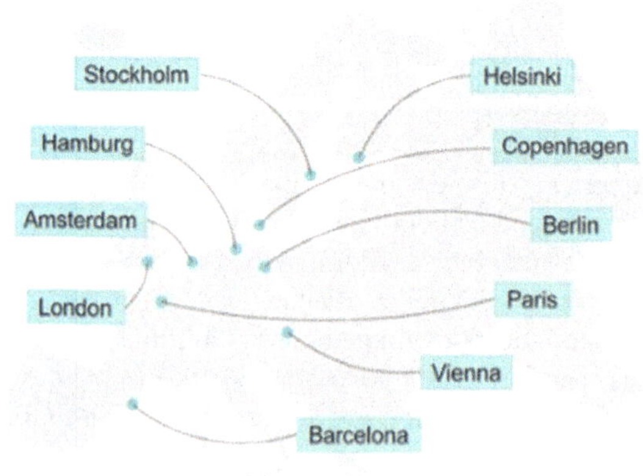

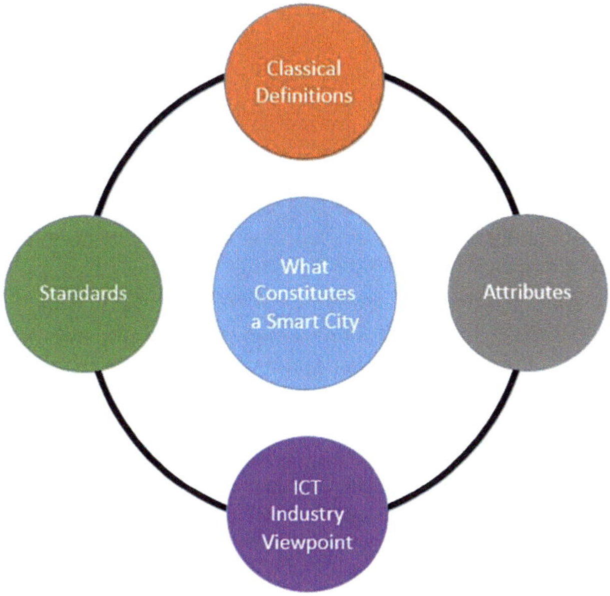

Fig. 3 What constitutes a smart city

The authors in [10] develop a case study for the "Padova smart city project" in Padova, Italy; the city is an IoT island proof-of-concept developed by private parties in collaboration with the municipality of Padova. In their paper the authors discuss about the fundamentals that were considered and implemented in order to realize the smart city and also the data collected in the first few operation days of the city. The main goal of the development of this city was to promote open data and ICT solutions in the public sector. The layout of this city consisted of the use of sensors planted on streetlights to collect and monitor environmental data such as CO level, air temperature and humidity, vibrations, noise, and so on, in order to efficiently and accurately manage the streetlighting by measuring the light intensity on each pole. A conceptual sketch of the Padova Smart City system architecture is given in Fig. 3.

3 Smart City Demarcation

As stated by the authors in [3], a smart city is a futuristic urban center, made safe, secure environmentally green, and efficient because all structures, such as water, power, transportation, and the likes, are designed and constructed using the latest technology such as sensors, electronics, and networks and interfaced with computerized systems made up of databases, tracking, and decision-making algorithms [32]. Moreover, the author in [4] cites four possible definitions of intelligent cities.

The first one is about the application of a wide range of electronics and digital application to communities and cities, which effectively work to combine the term – smart – with concepts of cyber, digital, wired, informational, or knowledge-based city. A second meaning is the use of information technology to positively transform the livelihood and work of people living in a given region. The third definition of "intelligent" is associated with the embedded information and communication technology in the city. The final definition is as a paradigm that brings ICTs and people together to enhance knowledge, learning, innovation, and problem-solving.

While stating that there is no clear definition of a smart city, the author in [5] says that governance, technology, communication, transport, infrastructure, people, economy, environment, natural resources, innovation, and quality of living are just some of the few components that factor in on the definition of a smart city. The author however goes on to explain there have been attempts to warrant a standardized definition to the term "smart city" by several Standards Developing Organizations (SDOs). One of the definitions of a smart sustainable city put forward by International Telecommunication Union (ITU) is:

"A smart sustainable city is an innovative city that uses information and communication technologies (ICTs) and other means to improve quality of life, efficiency of urban operation and services, and competitiveness, while ensuring that it meets the needs of present and future generations with respect to economic, social and environmental aspects."

The author in [5] concurs that there are quite a number of ways used to describe "smart city," each with its own valid reasons; the author also agrees that there is a need to develop a compressive definition which addresses all of these viewpoints and answer the question "What constitutes a smart city?" Consequently, in order to develop such a description, Kondepudi et al. [5] developed a detailed and precise analysis in the form of a blend of "top-down and bottom-up approaches." Figure 3 shows the approach followed by the authors so as to find out what constitutes a smart city.

On the other hand, Giffinger et al. [6] list four components that constitute a smart city; these are industry, education, participation, and technical infrastructure. However, this list has been extended in a recent study conducted by the Centre of Regional Science at the Vienna University of Technology for up to six components; these are smart economy, smart mobility, a smart environment, smart people, smart living, and smart governance [7]. Moreover, Lombardi et al. [8] have associated the six components with different aspects of urban life. Table 1 shows the summery of the related aspects.

4 A Connected City

The authors in [10] claim that there is no formal definition of the term "smart city"; however, the authors go on to say that the main goal of a smart city is to provide better and reliable services to the people at a minimum cost; this can

Table 1 Abbreviations used in this work

Abbreviation	Description
IoT	Internet of Things
SOM	Smart outdoor monitoring
ICT	Information communication technology
SDOs	Standards Developing Organizations
ITU	International Telecommunication Union
IPv4	Internet Protocol Version 4
IPv6	Internet Protocol Version 6
SSL	Secure Sockets Layer
HTTP	Hypertext Transfer Protocol
TCP	Transmission Control Protocol
CoAP	Constrained Application Protocol
UDP	User Datagram Protocol
6LoWPAN	IPv6 over Low-Power Wireless Personal Area Network
RPL	Recognition of prior learning
EU	European Union
V2V	Vehicle-to-vehicle
CoTEC	COoperative Traffic congestion detECtion
GPS	Global positioning system
RFID	Radio-frequency identification
NFC	Near-field communication
LAN	Local area network
WSN	Wireless sensor network
MTC	Machine-type communications
VoIP	Voice over Internet protocol
QoS	Quality of service
LTE	Long-Term Evolution
LTE-A	Long Term Evolution-Advanced
5G	Fifth generation network
4G	Fourth generation network
3G	Third generation network

be done by deploying the use of *urban IoT* [10]. It is a communication charnel that connects the entire urban area, easing up management and operation of public resources (transport and parking, lighting, surveillance and maintenance of public areas, preservation of cultural heritage, garbage collection, salubrity of hospitals, and school) and opening up a surplus of other possibilities. For instance, the data collected by sensors can be used to provide transparency to the people and hence leading to better management and the use of public resource [31]. It is therefore prudent to say that IoT is what connects our cities and ultimately turns an urban settlement into a smart city.

The authors in [9] state that IoT is a next-generation communication paradigm which envisions that the objects of everyday life are provided with communication capabilities, that is, equipped with microcontrollers, transceivers for digital communication, and suitable protocol stacks that will allow them to talk to each other and hence becoming an integral part of the Internet. Thus, the IoT purposes at making the Internet more immersive and universal. Moreover, by enabling access to "things" such as home appliances, surveillance cameras, monitoring sensors, actuators, displays, vehicles, and so on, the IoT open doors to the development of countless applications that make use of the enormous and variant data collected. Additionally, the systems developed are used to provide services to the general public and private and public organization. This paradigm is applied in so many other areas such as home automation, traffic management, elderly assistance, industrial automation, medical aids, automotive, mobile healthcare [37], intelligent energy management, and smart grids, among many others [11]. For the rest of this section, we look at how IoT influences selected services in a smart city to bring about advantages such as enhancing the quality of services offered to the people and increasing their quality of life while reducing the cost of operation for the city [12] (Fig. 4 and Table 2).

Structural Health of Buildings For proper maintenance of historic landmarks, such as old buildings in a city, it is imperative to carry out continued monitoring and restoration of structures. IoT allows for a continuous inspection of this buildings and a timely repair, hence saving time and money. IoT can facilitate the creation

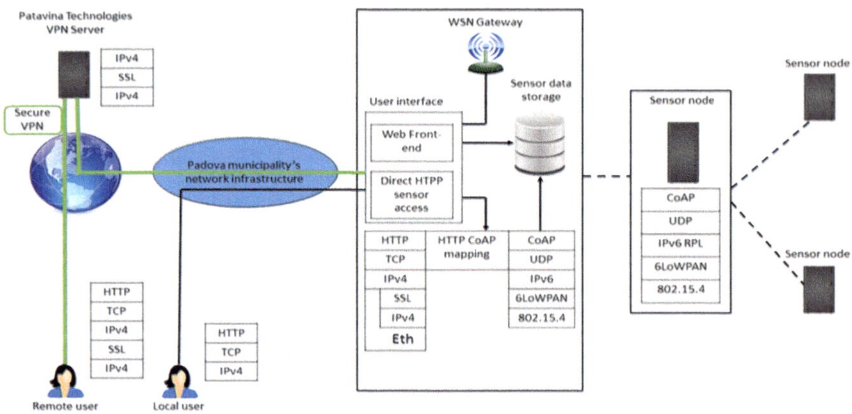

Fig. 4 System architecture of "Padova Smart City"

Table 2 Smart city components and its related aspects

Components of a smart city	Related aspect of urban life
Smart economy	Industry
Smart people	Education
Smart governance	E-democracy
Smart mobility	Logistics and infrastructures
Smart environment	Efficiency and sustainability
Smart living	Security and quality

of a database that consists of data collected by the sensors (e.g., vibration and deformation sensors, atmospheric agent sensors, temperature, and humidity sensors [13]) deployed on the buildings. The information in the database will be, for instance, the stress condition of the building, climatic condition around the building, the amount of pollution caused by the building, and the likes. This information will be available to the public, and most of all, the information can be used to predict the stability of the city to withstand natural phenomena such as earthquakes and tsunamis. The practical realization of this system, however, requires the investment of some initial capital, since there is a need to install sensors on the buildings and the surrounding environment and their interconnection to the database network [10].

Waste Management This is one major issue faced by most cities around the world, because of the high cost of operation and the environmental impact cause by dumping sites. However, if we indulge ICT in the waste management project, we will benefit more, for instance, the author in [14] talks about intelligent waste containers. The containers detect when it is full and communicate to the base and hence enable a timely collection and reduce operation cost; moreover, the containers also improve recycling quality. The containers are connected to a base station where an optimization software calculates an optimal route for the garbage collecting fleet.

Air Quality Global warming has brought about major environmental hazards, rising sea levels, melting ice caps in the polar regions, long dry seasons, and so on. Recently, many countries and organizations around the world decided to join forces to reduce global warming. The EU, among them, decided to reduce greenhouse gas emission by 20% by the year 2020, reduce energy consumption by 20% by the year 2020, and increase the use of renewable energy sources by 20% by the year 2020 [10]. To that extent, adoption of intelligent cities can greatly help in achieving this goal by monitoring the environmental condition in areas such as crowded places, parks, or fitness trails [15]. Moreover, the data collected by the air quality and pollution sensors can be stored in a database that is accessible to the public. This data can be helpful to people who conduct their activities outdoor, for instance, joggers can be provided with the ultimate route suitable for jogging.

Noise Monitoring Noise can be defined as a collection of sounds that is loud, irritating, and annoying to the listener. It can also cause hearing problems if it is loud enough. Many people experience noise indoors and outdoors [16]. A smart city can offer noise monitoring services to measure noise at specific areas at a given time [17]. Moreover, the sound-detecting services can be enhanced by means of a sound-detecting algorithm to decipher vandalism noises like braking glass and alerting a security department [36] and hence keeping the area safe for business developers. However, the deployment of sound detectors or environmental microphones is quite debatable because of the obvious reason of privacy [10].

Traffic Congestion The mystery of traffic congestion is one of the major puzzles faced by major cities around the world. Although traffic cameras have helped ease the congestion, low energy and widespread communication can greatly improve the efficiency of traffic monitoring. Traffic congestion can be realized in a smart city by deploying sensors along some major roads and using GPS capabilities installed in modern vehicles [18]. Moreover, the author in [19] proposes CoTEC (COoperative Traffic congestion detECtion), a system that uses vehicle-to-vehicle (V2V) communications to detect and control traffic without the need to deploy sensors. Nevertheless, the variant information collected by the sensors can be very useful to city authorities by detecting accident sites [39, 42, 44] and to citizens by helping them plan for an efficient route to the city.

City Energy Consumption In smart cities, it is also possible to measure the overall energy consumption of the city. This will give the authorities and citizens a clear analysis of how much power is being used in different sectors, for instance, public lighting, transportation, traffic lights, control cameras, heating/cooling of public buildings, and so on. In order to achieve this, power monitoring devices must be incorporated into the grid. This system can help the EU achieve their set goal to reduce power consumption by 20% in the coming years [10].

Smart Parking This system is made up of road sensors and smart displays, which are used to guide a driver to an available parking spot in a smart city [20, 46, 48]. There are a number of advantages brought about by this system, less time to locate a packing slot, leading to less CO emission, less traffic, and happier clients. Moreover, the use of short-range communication technology, such as *radio-frequency identification (RFID)* or near-field communication (NFC), can be used to model an electronic verification system for reserved and disable parking spaces and hence leading to better management of the parking lot [47].

Smart Lighting In order to reduce the overall city power consumption, streetlights have to be modeled to automatically adjust depending on the time of the day, weather conditions, and the presence of people. In a smart city, such systems can be incorporated using the already existing weather and sound sensors and hence reducing power consumption [34, 35]. The authors in [21] talk about the importance of a smart city in the modern world and go on to highlight that one key goal of a smart city is to reduce power consumption. Moreover, the main topic given by the authors is how to reach an interoperable smart lighting solution over the emerging M2M protocols.

Automation and Salubrity of Public Buildings One other important aspect of IoT in smart cities is the ability to monitor and control energy consumption and salubrity in public places such as schools, sport centers, and museums. This is done by the deployment of sensors that measure lights, temperature, and humidity in and around the buildings. The ability to control and regulate this parameter allows us to enhance the comfort of people in these buildings and, in some cases, increase productivity [22].

As evidently depicted in the presiding sections, communication in a smart city involves the exchange of big data from all the interconnected infrastructure elements. We therefore must have a network system in place that is efficient enough to handle such a massive movement of sensitive information.

5 Telecommunication in Smart Cities

Information and communication are key to the intelligent city of tomorrow. Communication, data exchange, and interconnectedness are the foundation for all other key subjects of a smart city: power supply, mobility, and public safety. In this section, we look at the most suitable channel of communication in a smart city. To start from the top, we first consider guided and unguided communication channels. Guided communication channel, or wired communication, is where the communication devices are directly connected to each other through a cable. Guided communication channels are commonly known as LAN. Some examples include coaxial cable, twisted pair wire, and fiber-optic cable. On the other hand, an unguided communication channel is a system where data is transferred in the form of a wave; this means that data does not follow a specific path. Moreover, information can be send all over the world. This system is commonly known as wireless communication; some examples are microwave, cellular radio, radio broadcast, and satellite. In a smart city paradigm, developers are looking for new ways to maximize their investment in communication networks while ensuring reliable and secure data transmission, in light of that, while both channels have a place in utility market applications, such as distribution mechanization, we are beginning to see an increase in the use of wireless systems network. The authors in [23] state that the use of wireless network as opposed to wired will cut down the installations and repairment cost.

Compared to fiber optics, for example, wireless communication systems are much easier to maintain. If a buried cable is damaged, it will take a lot of time, manpower, and money to repair. Nevertheless, wireless systems send a notification if any maintenance is required and can prevent the actual damage from occurring; however, in case maintenance is required, fewer personnel are needed to repair a wireless system compared to a wired system, consequently, saving time and money. Moreover, cables are priced by their length (miles, meters, feet), while wireless data radios' pricing includes ranges expressed in miles. Vandalism is one other major challenge facing the wired system. The transmission cables are usually made of expensive metal, such as copper; therefore, these cables are prone to be vandalized by thieves; a research done by Sidebottom et al. [23], on the theft of coper cables on the British railway network, show that there is an increased number of vandalized cables when the market price for copper goes high. High costs, difficult installations, copper theft, and more are just some of the factors that is driving investors to choose wireless communication over wired. In many industries including the military, there is evidence that wireless communication is a viable alternative. These technologies provide long-range, reliable, and affordable solutions. WSN technology can potentially save related companies millions of dollars in installation cost.

In the modern IoT connectivity landscape, 3G and 4G are the most engaging form of connectivity. These platforms offer extensive coverage, quite low deployment costs, high security levels, and simplicity in management [23]. However, there is one drawback, since the systems were designed for optimized broadband telecommunication; they do not support low-cost machine-type communications (MTC) – a fundamental aspect in the realization of a smart city. Consequently, we look at how 5G will improve its predecessors and move us closer to developing a smarter city.

Palattella et al. [25] provide an in-depth analysis about the potential of 5G technologies for the IoT. The authors state that 5G has an edge compared to 4G and 3G. The increased data rate, reduced end-to-end latency, and improved coverage equip 5G to handle even the most demanding and sophisticated application in IoT in relation to user satisfaction and the communication requirements. The development of 5G as a heterogeneous system, that is, the ability to facilitate the interconnection of a myriad of "things" with the Internet, puts it forward as a unified interconnected framework [25]. Moreover, fifth-generation wireless technologies will also provide very high-speed broadband wireless connectivity and offer various services such as documentation services, supporting fast electronic transactions such as e-payment, e-transaction, and the likes [25]. Voice over Internet protocol (VoIP) is a connection of devices over the Internet using their IP address, where voice data is send over the connection. VoIP is the fastest-growing communication technology [26]. The author in [27] says that compared to contemporary telephone networks, VoIP has two main advantages, improved bandwidth efficiency and the ability to facilitate the creation of new services that will be able to combine voice communication with other media and data applications. 5G mainly focuses on VoIP devices to warrant that users experience high level of call volume and data transmission [28]. 5G is specifically designed to provide quality of service to its users, by providing maximum bandwidth. Moreover, 5G has also been able to overcome some other network challenges such as latency, error rate, and uptime. Zhang et al. [29] say that 5G is expected to provide QoS by guaranteeing low latency to a wide spectrum of services, users, and applications with varied requirements.

The authors in [30] highlight some examples of 5G deployment infrastructure: integrated radio base station, centralized baseband, and radio-over-copper solutions. 5G networks have a limited communication range; therefore, a high number of cells need to be deployed so as to cover a large area. Consequently, at the beginning, the service providers need not abandon the use of the prevailing LTE and LTE-A-based infrastructure.

6 Conclusion

Smart cities are the future of urban development. Current advancement in technology, especially in medicine, is changing the lives of many people and increasing life expectancy; this means that there is a continuous population increase. Figure 1 shows that there has been a constant increase in the number of people living in urban areas since 1950. Smart cities provide a sure solution to some of the challenges brought about by overpopulation: waste management problems, scarcity of resources, air pollution, human health concerns, traffic congestions, and failing and aging infrastructures, among others. However, intelligent cities are only as good as the interconnection of devices used to collect and transmit information such as sensors in vehicles and streetlights and health monitoring devices in patients. Consequently, there is a large amount of information that is collected and analyzed every day in a smart city; it is imperative that this information be transmitted as fast as possible and, most importantly, as secure as possible. Internet of Things (IoT) is an ecosystem of interconnected physical devices, accessible through the Internet; this ecosystem will have a huge impact in the communication system of a smart city. As discussed in this study, wireless communication is highly preferred as a means of communication in a smart city topology as opposed to wired, with 5G possessing most the ideal requirements.

References

1. Maria, S., & Echsner-Rasmussen, N. (2015). Smart cities around the world. *Geoforum Perspektiv, 14*(25), 61–67.
2. Nations, United. (2015). *World urbanization prospects: The 2014 revision, highlights*. New York: Department of economic and social affairs. Population Division.
3. Braverman, J., Taylor, J., Todosow, H., & von Wimmersperg, U. (2000). *The vision of a smart city*. Upton, NY: Brookhaven National Lab..
4. Komninos, N., & Sefertzi, E. (2013). *Intelligent cities: innovation, knowledge systems and digital spaces*. London: Routledge.
5. Sekhar, K., & Kondepudi, R. (2015). What constitutes a smart city? In *Handbook of research on social, economic, and environmental sustainability in the development of smart cities* (pp. 1–25). Hershey, PA: IGI Global.
6. Giffinger, R., Fertner, C., Kramar, H., Kalasek, R., Pichler-Milanović, N., & Meijers, E. (2007). *Smart cities: ranking of European medium-sized cities*. Vienna: Centre of Regional Science.
7. Giffinger, R., & Gudrun, H. (2010). Smart cities ranking: an effective instrument for the positioning of cities? *ACE Architecture, City and Environment, 4*(12), 7–25.
8. Lombardi, P., Giordano, S., Farouh, H., & Yousef, W. (2012). Modelling the smart city performance. *Innovation: The European Journal of Social Science Research, 25*(2), 137–149.
9. Atzori, L., Iera, A., & Morabito, G. (2010). The internet of things: A survey. *Computer Networks, 54*(15), 2787–2805.
10. Andrea, Z., Bui, N., Castellani, A., Vangelista, L., & Zorzi, M. (2014). Internet of things for smart cities. *IEEE Internet of Things Journal., 1*(1), 22–32.
11. Bellavista, P., Cardone, G., Corradi, A., & Foschini, L. (2013). Convergence of MANET and WSN in IoT urban scenarios. *IEEE Sensors Journal, 13*(10), 3558–3567.
12. Dohler, M., Vilajosana, I., Vilajosana, X., & Llosa, J. (2011) Smart cities: An action plan. In *Proceedings of Barcelona Smart Cities Congress*, Barcelona, Spain, 2011, p. 6.
13. Lynch, J. P., & Kenneth, J. L. (2006). A summary review of wireless sensors and sensor networks for structural health monitoring. *Shock and Vibration Digest, 38*(2), 91–130.
14. Nuortio, T., Kytöjoki, J., Niska, H., & Bräysy, O. (2006). Improved route planning and scheduling of waste collection and transport. *Expert Systems with Applications, 30*(2), 223–232.
15. Al-Ali, A. R., Zualkernan, I., & Aloul, F. (2010). A mobile GPRS-sensors array for air pollution monitoring. *IEEE Sensors Journal., 10*(10), 1666–1671.
16. Eiman, K. (2010). Noisespy: A real-time mobile phone platform for urban noise monitoring and mapping. *Mobile Networks and Applications, 15*(4), 562–574.

17. Maisonneuve, N., Stevens, M., Niessen, M. E., Hanappe, P., & Steels, L. (2009). Citizen noise pollution monitoring. In *Proceeding of 10th Annual International Conference on Digital Government Research: Social Networks: Making Connections between Citizens, Data and Government* (pp. 96–103).
18. Li, X., Shu, W., Li, M., Huang, H.-Y., Luo, P.-E., & Wu, M.-Y. (2009). Performance evaluation of vehicle-based mobile sensor networks for traffic monitoring. *IEEE Transactions on Vehicular Technology, 58*(4), 1647–1653.
19. Ramon, B., Gozalvez, J., & Sanchez-Soriano, J. (2010). Road traffic congestion detection through cooperative vehicle-to-vehicle communications. In *Proceeding of Local Computer Networks (LCN), 2010 IEEE 35th Conference on* (pp. 606–612).
20. Lee, S., Yoon, D., & Ghosh, A. (2008). Intelligent parking lot application using wireless sensor networks. In *Proceeding of International Collaborative Technologies and Systems*, Chicago, May 19–23, (pp. 48–57).
21. Miguel, C., Jara, A., & Skarmeta, A.. (2013). Smart lighting solutions for smart cities. In *Proceeding of 2th IEEE International Conference on Advanced Information Networking and Applications Workshops (WAINA)*, (pp. 1374–1379).
22. Kastner, W., Neugschwandtner, G., Soucek, S., & Newmann, H. M. (2005). Communication systems for building automation and control. *Proceedings of the IEEE, 93*(6), 1178–1203.
23. Lemayian, J. P., Abdelhamid, S., Alturjman, S., Ever, E., & Al-Turjman, F. (2018). 5G in a Convergent internet of things Era: An overview. In *2018 IEEE International Conference on Communications Workshops (ICC Workshops).*
24. Rita, P. M., Dohler, M., Grieco, A., Rizzo, G., Torsner, J., Engel, T., & Ladid, L. (2016). Internet of things in the 5G era: Enablers, architecture, and business models. *IEEE Journal on Selected Areas in Communications, 34*(3), 510–527.
25. Dai, J., Bai, X., Yang, Z., Shen, Z., & Xuan, D., Perfalld: A pervasive fall detection system using mobile phones, Pervasive Computing and Communications Workshops (PERCOM Workshops). *2010 8th IEEE International Conference on*, pp. 292–297, 29 2010-April 12 2010.
26. Lin, Y.-B., & Chlamtac, I. (2001). *Wireless and mobile network architectures*. New York: Wiley.
27. Wei, W., Liew, S. C., & Li, V. O. K. (2005). Solutions to performance problems in VoIP over a 802.11 wireless LAN. *IEEE Transactions on Vehicular Technology, 54*(1), 366–384.
28. Sapakal, R., & Kadam, S. (2013). 5G mobile technology. *International Journal of Advanced Research in Computer Engineering & Technology (IJARCET), 2*, 568–571.
29. Xi, Z., Cheng, W., & Zhang, H. (2014). Heterogeneous statistical QoS provisioning over 5G mobile wireless networks. *IEEE Network, 28*(6), 46–53.
30. Maisonneuve, N., Stevens, M., Niessen, M. E., Steels, L., Allan, R., Frstner, U., & Salomons, W. (2009). Noise tube: Measuring and mapping noise pollution with mobile phones. In *Information Technologies in Environmental Engineering ser. Environmental Science and Engineering* (pp. 215–228). Berlin Heidelberg: Springer.
31. Al-Turjman, F., & Alturjman, S. (2018). Confidential smart-sensing framework in the IoT Era. *The Springer Journal of Supercomputing, 74*, 5187. https://doi.org/10.1007/s11227-018-2524-1.
32. Al-Turjman, F., & Alturjman, S. (2018). 5G/IoT-enabled UAVs for multimedia delivery in industry-oriented applications. *Springer's Multimedia Tools and Applications Journal*. https://doi.org/10.1007/s11042-018-6288-7.
33. Al-Turjman, F., & Abdulsalam, A. (2018). Smart-grid and solar energy harvesting in the IoT Era: An overview. *Wiley's Concurrency and Computation: Practice and Experience*. https://doi.org/10.1002/cpe.4896.
34. Demir, S., & Al-Turjman, F. (2018). Energy scavenging methods for WBAN applications: A review. *IEEE Sensors Journal, 18*(16), 6477–6488.
35. Alabady, S., & Al-Turjman, F. (2018). A novel approach for error detection and correction for efficient energy in wireless networks. *Springer Multimedia Tools and Applications*. https://doi.org/10.1007/s11042-018-6282-0.

36. Alabady, S., Al-Turjman, F., & Din, S. (2018). A novel security model for cooperative virtual networks in the IoT era. *Springer International Journal of Parallel Programming*. https://doi.org/10.1007/s10766-018-0580-z.
37. Al-Turjman, F., & Alturjman, S. (2018). Context-sensitive access in industrial internet of things (IIoT) healthcare applications. *IEEE Transactions on Industrial Informatics, 14*(6), 2736–2744.
38. Al-Turjman, F. (2017). 5G-enabled devices and smart-spaces in social-IoT: An overview. *Elsevier Future Generation Computer Systems*. https://doi.org/10.1016/j.future.2017.11.035.
39. Al-Turjman, F. (2017). A rational data delivery framework for disaster-inspired internet of nano-things (IoNT) in practice. *Springer Cluster Computing*. https://doi.org/10.1007/s10586-017-1357-7.
40. Al-Turjman, F. (2018). Modelling green femtocells in smart-grids. *Springer Mobile Networks and Applications, 23*(4), 940–955.
41. Al-Turjman, F. (2018). Optimized hexagon-based deployment for large-scale ubiquitous sensor networks. *Springer's Journal of Network and Systems Management, 26*(2), 255–283.
42. Al-Turjman, F. (2017). Cognitive routing protocol for disaster-inspired internet of things. *Elsevier Future Generation Computer Systems*. https://doi.org/10.1016/j.future.2017.03.014.
43. Al-Turjman, F. (2017). Cognitive-node architecture and a deployment strategy for the future sensor networks. *Springer Mobile Networks and Applications*. https://doi.org/10.1007/s11036-017-0891-0.
44. Yatbaz, H., Cinar, B., Gokdemir, A., Ever, E., Al-Turjman, F., Nguyen, H., & Yazici, A. (2018). Hybrid approach for disaster recovery using P2P communications in android. In *Proceeding of the IEEE Local Computer Networks (LCN)*, Chicago, USA, (Accepted).
45. Qadir, Z., Al-Turjman, F., Tafadzwa, V., & Rashid, H. (2018). Smart solar microgrid using zigbee and related security challenges. In *International Conference on Research in Education and Science (ICRES)*, Marmaris, Turkey, April, (Accepted).
46. Qadir, Z., Al-Turjman, F., & Nesimoglu, T. (2018). ZIGBEE based time and energy efficient smart parking system using IOT. In *International Conference on Research in Education and Science (ICRES)*, Marmaris, Turkey, April, (Accepted).
47. Campioni, F., Choudhury, S. & Al-Turjman, F. (2018). Readers scheduling for RFID networks in the IoT Era. In *Proceeding of the IEEE International Conf. on Communications (ICC)*, Kansas City, MO, USA, (Accepted).
48. Kizilkaya, B., Caglar, M., Al-Turjman, F., & Ever, E. (2018). An intelligent car park management system: Hierarchical placement algorithm based on nearest location. In *Proceeding of the IEEE Int. Conf. on Advanced Information Networking and Applications*, Cracow, Poland, (Accepted).

Index

F. Al-Turjman (ed.), *Artificial Intelligence in IoT*,
Transactions on Computational Science and Computational Intelligence,
https://doi.org/10.1007/978-3-030-04110-6

T

Printed by Printforce, the Netherlands